EDITED BY MICHAEL RUSE
AND CHRISTOPHER A. PYNES

THE STEM CELL CONTROVERSY

DEBATING THE ISSUES

SECOND EDITION

Prometheus Books
59 John Glenn Drive
Amherst, New York 14228-2197

Published 2006 by Prometheus Books

Inquiries should be addressed to
Prometheus Books
59 John Glenn Drive
Amherst, New York 14228–2197
VOICE: 716–691–0133, ext. 207
FAX: 716–564–2711
WWW.PROMETHEUSBOOKS.COM

10 09 08 07 06   5 4 3 2 1

Library of Congress Cataloging-in-Publication Data

The stem cell controversy: debating the issues / edited by Michael Ruse and
Christopher A. Pynes. — 2nd. ed.
p. cm. (Contemporary Issues series)
Includes bibliographical references.
ISBN 13: 978–1–59102–404–0 (alk. paper)
ISBN 10: 1–59102–404–8 (alk. paper)
1. Stem cells—Research—Moral and ethical aspects. I. Ruse, Michael.
II. Pynes, Christopher A., 1970–

QH587.S723 2006
174.2'8—dc21

2006013373

Printed in the United States on acid-free paper

THE STEM CELL CONTROVERSY

Contemporary Issues

Series Editors: Robert M. Baird & Stuart E. Rosenbaum

Volumes edited by Robert M. Baird and Stuart E. Rosenbaum unless otherwise noted.

Affirmative Action: Social Justice or Reverse Discrimination?
edited by Francis J. Beckwith and Todd E. Jones

Animal Experimentation: The Moral Issues

Are You Politically Correct?: Debating America's Cultural Standards
edited by Francis J. Beckwith and Michael E. Bauman

Caring for the Dying: Critical Issues at the Edge of Life

Civil Liberties vs. National Security in a Post-9/11 World
edited by M. Katherine B. Darmer, Robert M. Baird, and Stuart E. Rosenbaum

Cloning: Responsible Science or Technomadness?
edited by Michael Ruse and Aryne Sheppard

Cyberethics: Social and Moral Issues in the Computer Age
edited by Robert M. Baird, Reagan Ramsower, and Stuart E. Rosenbaum

Drugs: Should We Legalize, Decriminalize, or Deregulate?
edited by Jeffrey A. Schaler

The Ethics of Abortion: Pro-Life vs. Pro-Choice
(Third edition)

The Ethics of Organ Transplants: The Current Debate
edited by Arthur L. Caplan and Daniel H. Coelho

Euthanasia: The Moral Issues

Food for Thought: The Debate over Eating Meat
edited by Steve F. Sapontzis

Generations Apart: Xers vs. Boomers vs. the Elderly
edited by Richard D. Thau and Jay S. Heflin

Genetically Modified Foods: Debating Biotechnology
edited by Michael Ruse and David Castle

The Gun Control Debate: You Decide
(Second edition) edited by Lee Nisbet

Hatred, Bigotry, and Prejudice: Definitions, Causes, and Solutions

Homosexuality: Debating the Issues
edited by Robert M. Baird and M. Katherine Baird

Immigration: Debating the Issues
edited by Nicholas Capaldi

The Media and Morality
edited by Robert M. Baird, William E. Loges, and Stuart E. Rosenbaum

Morality and the Law

Pornography: Private Right or Public Menace? (Revised edition)

Punishment and the Death Penalty: The Current Debate

Same-Sex Marriage: The Moral and Legal Debate (Second edition)

The Science Wars: Debating Scientific Knowledge and Technology
edited by Keith Parsons

Sexual Harassment: Confrontations and Decisions
(Revised edition) edited by Edmund Wall

Should Parents Be Licensed? Debating the Issues
edited by Peg Tittle

Smoking: Who Has the Right?
edited by Jeffrey A. Schaler and Magda E. Schaler

The Stem Cell Controversy: Debating the Issues (Second edition)
edited by Michael Ruse and Christopher A. Pynes

Suicide: Right or Wrong? (Second edition)
edited by John Donnelly

Contents

PART 2. MEDICAL CURES AND PROMISES

PART 3. MORAL ISSUES

PART 4. RELIGIOUS ISSUES

PART 5. POLICY ISSUES

President George W. Bush

on Stem Cell Research

Good evening. I appreciate you giving me a few minutes of your time tonight so I can discuss with you a complex and difficult issue, an issue that is one of the most profound of our time.

The issue of research involving stem cells derived from human embryos is increasingly the subject of a national debate and dinner table discussions. The issue is confronted every day in laboratories as scientists ponder the ethical ramifications of their work. It is agonized over by parents and many couples as they try to have children, or to save children already born.

The issue is debated within the church, with people of different faiths, even many of the same faith coming to different conclusions. Many people are finding that the more they know about stem cell research, the less certain they are about the right ethical and moral conclusions.

My administration must decide whether to allow federal funds, your tax dollars, to be used for scientific research on stem cells derived from human embryos. A large number of these embryos already exist. They are the product of a process called in vitro fertil-

White House news release, August 9, 2001.

ization, which helps so many couples conceive children. When doctors match sperm and egg to create life outside the womb, they usually produce more embryos than are planted in the mother. Once a couple successfully has children, or if they are unsuccessful, the additional embryos remain frozen in laboratories.

Some will not survive during long storage; others are destroyed. A number have been donated to science and used to create privately funded stem cell lines. And a few have been implanted in an adoptive mother and born, and are today healthy children.

Based on preliminary work that has been privately funded, scientists believe further research using stem cells offers great promise that could help improve the lives of those who suffer from many terrible diseases—from juvenile diabetes to Alzheimer's, from Parkinson's to spinal cord injuries. And while scientists admit they are not yet certain, they believe stem cells derived from embryos have unique potential.

You should also know that stem cells can be derived from sources other than embryos—from adult cells, from umbilical cords that are discarded after babies are born, from human placenta. And many scientists feel research on these type of stem cells is also promising. Many patients suffering from a range of diseases are already being helped with treatments developed from adult stem cells.

However, most scientists, at least today, believe that research on embryonic stem cells offer the most promise because these cells have the potential to develop in all of the tissues in the body.

Scientists further believe that rapid progress in this research will come only with federal funds. Federal dollars help attract the best and brightest scientists. They ensure new discoveries are widely shared at the largest number of research facilities and that the research is directed toward the greatest public good.

The United States has a long and proud record of leading the world toward advances in science and medicine that improve human life. And the United States has a long and proud record of upholding the highest standards of ethics as we expand the limits of science and knowledge. Research on embryonic stem cells raises profound ethical questions, because extracting the stem cell destroys the embryo, and thus destroys its potential for life. Like a snowflake, each of these embryos is unique, with the unique genetic potential of an individual human being.

As I thought through this issue, I kept returning to two fundamental questions: First, are these frozen embryos human life, and therefore, something precious to be protected? And second, if they're going to be destroyed anyway, shouldn't they be used for a greater good, for research that has the potential to save and improve other lives?

I've asked those questions and others of scientists, scholars, bioethicists, religious leaders, doctors, researchers, members of Congress, my Cabinet, and my friends. I have read heartfelt letters from many Americans. I have given this issue a great deal of thought, prayer and considerable reflection. And I have found widespread disagreement.

On the first issue, are these embryos human life—well, one researcher told me he believes this five-day-old cluster of cells is not an embryo, not yet an individual, but a pre-embryo. He argued that it has the potential for life, but it is not a life because it cannot develop on its own.

An ethicist dismissed that as a callous attempt at rationalization. Make no mistake, he told me, that cluster of cells is the same way you and I, and all the rest of us, started our lives. One goes with a heavy heart if we use these, he said, because we are dealing with the seeds of the next generation.

And to the other crucial question, if these are going to be destroyed anyway, why not use them for good purpose—I also found different answers. Many argue these embryos are by-products of a process that helps create life, and we should allow couples to donate them to science so they can be used for good purpose instead of wasting their potential. Others will argue there's no such thing as excess life, and the fact that a living being is going to die does not justify experimenting on it or exploiting it as a natural resource.

At its core, this issue forces us to confront fundamental questions about the beginnings of life and the ends of science. It lies at a difficult moral intersection, juxtaposing the need to protect life in all its phases with the prospect of saving and improving life in all its stages.

As the discoveries of modern science create tremendous hope, they also lay vast ethical mine fields. As the genius of science extends the horizons of what we can do, we increasingly confront complex questions about what we should do. We have arrived at that brave new world that seemed so distant in 1932, when Aldous Huxley wrote about human beings created in test tubes in what he called a "hatchery."

In recent weeks, we learned that scientists have created human embryos in test tubes solely to experiment on them. This is deeply troubling, and a warning sign that should prompt all of us to think through these issues very carefully.

Embryonic stem cell research is at the leading edge of a series of moral hazards. The initial stem cell researcher was at first reluctant to begin his research, fearing it might be used for human cloning. Scientists have already cloned a sheep. Researchers are telling us the next step could be to clone human beings to create individual designer stem cells, essentially to grow another you, to be available in case you need another heart or lung or liver.

I strongly oppose human cloning, as do most Americans. We recoil at the idea of growing human beings for spare body parts, or creating life for our convenience. And while we must devote enormous energy to conquering disease, it is equally important that we pay attention to the moral concerns raised by the new frontier of human embryo stem cell research. Even the most noble ends do not justify any means.

My position on these issues is shaped by deeply held beliefs. I'm a strong supporter of science and technology, and believe they have the potential for incredible good—to improve lives, to save life, to conquer disease. Research offers hope that millions of our loved ones may be cured of a disease and rid of their suffering. I have friends whose children suffer from juvenile diabetes. Nancy Reagan has written me about President Reagan's struggle with Alzheimer's. My own family has confronted the tragedy of childhood leukemia. And, like all Americans, I have great hope for cures.

I also believe human life is a sacred gift from our Creator. I worry about a culture that devalues life, and believe as your President I have an important obligation to foster and encourage respect for life in America and throughout the world. And while we're all hopeful about the potential of this research, no one can be certain that the science will live up to the hope it has generated.

Eight years ago, scientists believed fetal tissue research offered great hope for cures and treatments—yet, the progress to date has not lived up to its initial expectations. Embryonic stem cell research offers both great promise and great peril. So I have decided we must proceed with great care.

As a result of private research, more than sixty genetically diverse stem cell lines already exist. They were created from embryos that have already been destroyed, and they have the ability to regenerate themselves indefinitely, creating ongoing opportunities for research. I have concluded that we should allow federal funds to be used for research on these existing stem cell lines, where the life and death decision has already been made.

Leading scientists tell me research on these sixty lines has great promise that could lead to breakthrough therapies and cures. This allows us to explore the promise and potential of stem cell research without crossing a fundamental moral line, by providing taxpayer funding that would sanction or encourage further destruction of human embryos that have at least the potential for life.

I also believe that great scientific progress can be made through aggressive federal funding of research on umbilical cord, placenta, adult, and animal stem cells which do not involve the same moral dilemma. This year, your government will spend $250 million on this important research.

I will also name a president's council to monitor stem cell research, to recommend appropriate guidelines and regulations, and to consider all of the medical and ethical ramifications of biomedical innovation. This council will consist of leading scientists, doctors, ethicists, lawyers, theologians, and others, and will be chaired by Dr. Leon Kass, a leading biomedical ethicist from the University of Chicago.

This council will keep us apprised of new developments and give our nation a forum to continue to discuss and evaluate these important issues. As we go forward, I hope we will always be guided by both intellect and heart, by both our capabilities and our conscience.

I have made this decision with great care, and I pray it is the right one.

Thank you for listening. Good night, and God bless America.

Editors' General Introduction

Michael Ruse and Christopher A. Pynes

Parkinson's disease is a truly dreadful ailment that affects one and a half million Americans. It starts with the death of cells in the brain. This triggers a failure to produce dopamine, a substance needed to transmit information between nerve cells. The end result is the tremors—uncontrollable shaking of the hands—and eventually total rigidity. As the sad story of the actor Michael J. Fox shows only too well, although Parkinson's is mainly an affliction of the aged, it is something that can come down upon the young and is no respecter of status or achievement. There are some ways to slow its course, notably with drugs that produce dopamine, but these are temporary measures and eventually lose their force. Today, however, there is a whole new avenue of exciting research, an avenue that promises the possibility of complete and lasting relief from this disease. Certain cells of the body, so-called stem cells, have the ability to multiply and to make anew all of the parts of the organism, including the parts of the brain that have ceased to function in those cursed by Parkinson's. With the aim of inducing the brain itself again to produce dopamine, one might therefore think of grafting or injecting stem cells into sufferers. Research is still very much incomplete, but already there are some dramatic successes, with cases of people enjoying at least a 50 percent remission in their symptoms. And these are lasting, in some cases for ten years or more, with the grafted cells still doing their work.

The way forward is surely obvious. Pour funds into stem cell research. Encourage researchers to turn to its problems. Welcome the interest of the medical profession in such a potentially powerful new way of alleviating disease and misery—for Parkinson's is but one disease that might be cured or moderated by stem cell therapy. Rejoice that human ingenuity and compassion promises to do so much for so many. Unfortunately, things are not this easy or problem-free. Stem cells do not grow on trees. They are obtained from human beings and—fairly obviously really, when you think about it—although there are various possible sources (like the blood from the umbilical cord of newborns), the most ready font is the very, very young, who have not yet done much developing. It is they whose cells still have the greatest potential to transform into the many different parts of the functioning organism. But, of course, the truly very, very young are early embryos, and this at once raises massive ethical and religious problems, not to mention those of social and economic policy.

To put things in context, it can take six or more embryos to produce enough cells to treat a Parkinson's sufferer successfully. Six embryos that have been produced artificially in order to harvest their cells, or six embryos that have been conceived naturally and then aborted, whether or not the abortion was done with the deliberate end of harvesting stem cells. But is it right to treat human beings in this way, or—if you cavil at calling the early embryo a human being—is it right to treat potential human beings in this way? It is this issue that spurred President Clinton in late 1998 to turn for advice to his National Bioethics Advisory Commission (NBAC). It is this issue that President Bush addressed in his talk to the nation on August 9, 2001, as he sought to find a compromise—a middle way—between those for whom stem cell research and therapy is but one more weapon in our fight against ill health and disease, and those for whom any use of embryos, for whatever reason, is absolutely abhorrent and to be forbidden. And it is this issue that we address through the collection of readings in the volume you now hold in your hand. We wanted to give you, not the answers, for (even if we could) it is ultimately for you to decide these things for yourself, but at least some of the tools that you might use in order to come to the answers that you find full and satisfying. We start with the science, move then to the potential medical applications, look at both ethical and religious issues, and conclude with some writings about policy and the possible ways forward.

This is an explosive issue and going to get more so. The deliberate destruction of embryos—abortion—has split families, societies, nations. Thus far, one of the most important elements in the controversy has been the fact that, with obvious exceptions, the act of abortion usually occurs in a haze of moral disapproval. Or, at least, in circumstances and for reasons in which those who want to find moral objections can succeed in their quest. Unwanted pregnancies do not occur by chance but through carelessness or the like. Why then should a human or potential human be destroyed just to mop up after someone has failed to exercise restraint or caution? But the stem cell issue brings in another dimension entirely. If individuals fall sick, it is usually not their fault. Why then should they be denied the opportunity of cure if already there is an abundant supply of the needed material anyway—the millions of aborted fetuses that are already discarded every year? By what right should those who suffer be denied cure by the wishes or demands of those whose moral or religious convictions are not shared by the rest of us?

On these issues we have striven deliberately, not to be neutral in the sense of uncaring, but to be balanced in the sense of letting people of very different beliefs and persuasions all have their say. Even the science itself is not without controversy, for (as you will see) there are debates about the possible sources of stem cells. Potential applications raise even more questions and disagreements, hopes and false dreams. And then, as we move to ethical or moral issues, conflict really escalates, as people draw diametrically opposed conclusions about right courses of action. Expectedly, religion does little to calm these conflicts, and between adherents of different faiths there is at times stark opposition. Which means that, at the end, policymakers have to pick up the pieces and to make recommendations, to enact policies, and to try to enforce and to follow them. Hence the kind of proposal put forward by President Bush—to allow some stem cell research from already-existing lines, but not to allow (certainly not to fund by the government) the deliberate production of new sources of such cells.

We have chosen readings that are accessible to the nonspecialist, and to further aid comprehension we have added a glossary of unfamiliar terms and concepts. Some of the ideas in the section on ethics may be a little unfamiliar, especially if you have never taken a course in philosophy—unfamiliar, but not really that complicated. Just remember that people who deal with moral issues generally try to relate their thinking to a background theory or world picture. Thus, for instance, the Christian dealing with issues of world poverty would try to understand the issues in terms of Jesus' preaching in the Sermon on the Mount, as well as through the parables (like that of the Good Samaritan), together with the writings of great Christian thinkers from Saint Paul right down to our own day. Those who take a more secular approach to morality—that is, those who try to understand right and wrong without direct reference to a deity (although they may in fact themselves be deeply religious)—tend to fall into one of two camps. Either they are "deontologists"—judging matters by intention—or they are "teleologists" (or, to use another name for the latter, "consequentialists")—judging matters by results. Of course, often intentions and results come to the same thing—a person who sets out to do something is more likely to achieve the end than one who has no such aims—but intentions and results can come apart, and then real differences can emerge. Particularly the deontologist will stress the overall attitude that one has to people and their status and needs, whereas the consequentialist will stress the end result, downplaying the ways used to achieve that end.

The greatest of the deontologists was the late-eighteenth-century German philosopher Immanuel Kant, who argued that acts should be judged by what he called the "Categorical Imperative." The easiest formulation of this rule is that one should always treat people for their own sake, as ends in themselves, and never as means to some other good or end. Unless the victim in some way benefits as well, making an example of someone just for the general good is never justifiable. The greatest of the consequentialists were the British utilitarians, notably in the nineteenth century John Stuart Mill, who argued that one judges right actions by the results, especially the extent to which such actions maximize happiness and minimize unhappiness. One has to balance the benefits of one person or group against the benefits of other persons or groups. You might think that a Kantian would automatically come out against stem cell research, for surely in such an enterprise you are treating the embryo as a means to the ends of others. You would not balance the medical benefits of Nazi human experimentations against the suffering of the innocents on whom Dr. Mengele and other medical ghouls operated. Conversely, you might think that a utilitarian would automatically come out in favour of stem cell research, for the happiness of the sufferer will be eased and relieved and this will outweigh the ill or harm to the nonaware embryo. Most of us dislike the vivisection of dumb animals, but we swallow our scruples and let it happen. But, as you will learn, although there is undoubted truth in this expectation—Kantians do tend to be more conservative on moral issues, and utilitarians do tend to be more liberal about these things—matters are not entirely simple. Is the fetus really a human being, and if not, then does the Categorical Imperative apply to it? Conversely, is the happiness of the relieved sufferer enough to outweigh the potential happiness of the fetus? And in any case, what about the happiness of those people concerned for and involved with the embryo?

These are some of the questions you will see asked, and as you will see, different people come to different conclusions. Your conclusions must be your conclusions. What we hope is that this collection will help you to make them. For decisions must be made. As the great French philosopher Jean-Paul Sartre once said, not to chose a course of action is in fact to chose another course of action. If you do not think about these issues, then implicitly you are letting others do the thinking for you, and their conclusions are your conclusions, and their actions are your actions. And with such an important matter as stem cell research, this should not be so. As the great Anglican divine and poet John Donne once said: "No man is an island." We are all part of society and what happens to one of us happens to all of us. So join with us in reading what we have collected, and in thinking and reasoning about these very important issues—important for us and for our fellow citizens and indeed for all members of the human species.

INTRODUCTION TO THE SECOND EDITION

In this second edition, we have kept the original papers from the first edition intact, while adding what we feel is a useful addition of papers that helps fill out the stem cell debate. In this second edition, we have added more detailed introductions that give brief summaries of each of the works contained in the five sections, and we offer some questions to help guide readers as they move through the different articles. The addition of summaries and questions should give readers some idea as to how to think about the stem cell debate and questions at the forefront of science and ethics today. We hope that you enjoy this second edition of *The Stem Cell Controversy*.

Part 1
The Science of Stem Cells

Editors' Introduction

To begin at the beginning, we need to know something about the science. All living organisms are made of cells, some being but one cell and others (including the higher animals) of many different cells, each doing its allotted task or function. In sexual organisms like humans, the male and the female each contribute a special sex cell (the sperm and the ovum, respectively) that fuse together (into the zygote) that then divides and multiplies as the individual grows and develops. The early cells are known as stem cells, meaning that they can go on dividing indefinitely and that they have the potential to make all of the specialized cells needed in the body. We start our collection with a primer put out by the National Institutes of Health in May of 2000 that gives you the basics of the science, showing how stem cells arise and what their fates and potentialities hold. This will give you the basic science and much of the needed technical terminology, although you should then supplement this with the next reading, a specially commissioned piece by the historian and philosopher of science Jane Maienschein, who digs into the past showing that concepts and understanding only evolve gradually, that terminology does matter, and that it is important to know precisely what you are arguing about before you plunge into heated discussion. Sometimes people object that scholars, philosophers particularly, are overly concerned with language when they should be getting on with the facts, and there is perhaps some truth in this complaint. But words do matter, and a lot of heat later can be avoided by a little light now.

As you already saw in the president's address and in our general introduction, the really hard issues—moral and religious—come from the use of fetuses as a source of stem cells. Obviously if one could avoid or minimize the use of early embryonic material, much of the controversy would die and be avoided. A matter of some considerable interest therefore is whether one can harvest stem cells from older organisms—that is, from grown humans who would not be destroyed in the process of stem cell recovery and who could give their informed consent to the donation. At the moment, however, it is a hotly contested issue as to whether it is possible to use adults as a source for all of the stem cells that seem useful and needed. To give you some understanding of thc science, we present two readings, one somewhat pessimistic and another (specially commissioned from Sidney Houff of the Loyola University Medical Center in Chicago) that thinks the prospects are rather more promising. Then finally we present a fascinating news report by the *Washington Post* science reporter Rick Weiss. He discusses the possibility of stimulating cells to reproduce themselves and obtaining stem cells in that fashion. We see at once how important are some of Maienschein's concerns about language and definitions. These artificially stimulated cells—"parthenots"—are not like normal embryos, and some would argue that therefore they have no rights and raise no special moral and religious questions. Others would argue precisely the opposite. So much is going to depend on precisely what you mean by "normal," "potential," "embryo," "human," and other related terms.

QUESTIONS TO THINK ABOUT

(1) What exactly is a stem cell? How may different types of stem cells are there?

(2) Does a plurality of stem cell types change the moral debate in any important ways?
(3) How do language and names cause us to view different situations? For example, if we had a fertilized egg that we named William Smith Jr., would the act of naming it change the kind of relation we have with the cell?
(4) If all types of stem cells were equally useful in science, then would it matter where the cells came from—adults or fetuses or Petri dishes?
(5) Does parthenogenesis (virgin birth) in lower mammals and reptiles and recreated in the lab with human cells have any implications for the Christian belief in a virgin birth of Christ?

1
Stem Cells

A Primer

National Institutes of Health

This primer presents background information on stem cells. It includes an explanation of what stem cells are; what pluripotent stem cells are; how pluripotent stem cells are derived; why pluripotent stem cells are important to science; why they hold such great promise for advances in health care; and what adult stem cells are.

Recent published reports on the isolation and successful culturing of the first human pluripotent stem cell lines have generated great excitement and have brought biomedical research to the edge of a new frontier. The development of these human pluripotent stem cell lines deserves close scientific examination, evaluation of the promise for new therapies, and prevention strategies, and open discussion of the ethical issues.

In order to understand the importance of this discovery as well as the related scientific, medical, and ethical issues, it is absolutely essential to first clarify the terms and definitions.

From the National Institutes of Health, May 2000.

DEFINITIONS

DNA—abbreviation for deoxyribonucleic acid which makes up genes.

Gene—a functional unit of heredity which is a segment of DNA located in a specific site on a chromosome. A gene directs the formation of an enzyme or other protein.

Somatic cell—cell of the body other than egg or sperm.

Somatic cell nuclear transfer—the transfer of a cell nucleus from a somatic cell into an egg from which the nucleus has been removed.

Stem cells—cells that have the ability to divide for indefinite periods in culture and to give rise to specialized cells.

Pluripotent—capable of giving rise to most tissues of an organism.

Totipotent—having unlimited capability. Totipotent cells have the capacity to specialize into extraembryonic membranes and tissues, the embryo, and all postembryonic tissues and organs.

WHAT IS A STEM CELL?

Stem cells have the ability to divide for indefinite periods in culture and to give rise to specialized cells. They are best described in the context of normal human development. Human development begins when a sperm fertilizes an egg and creates a single cell that has the potential to form an entire organism. This fertilized egg is *totipotent*, meaning that its potential is total. In the first hours after fertilization, this cell divides into identical totipotent cells. This means that either one of these cells, if placed into a woman's uterus, has the potential to develop into a fetus. In fact, identical twins develop when two totipotent cells separate and develop into two individual, genetically identical human beings. Approximately four days after fertilization and after several cycles of cell division, these totipotent cells begin to specialize, forming a hollow sphere of cells, called a blastocyst. The blastocyst has an outer layer of cells and inside the hollow sphere, there is a cluster of cells called the inner cell mass.

The outer layer of cells will go on to form the placenta and other supporting tissues needed for fetal development in the uterus. The inner cell mass cells will go on to form virtually all of the tissues of the human body. Although the inner cell mass cells can form virtually every type of cell found in the human body, they cannot form an organism because they are unable to give rise to the placenta and supporting tissues necessary for development in the human uterus. These inner cell mass cells are *pluripotent*—they can give rise to many types of cells but not all types of cells necessary for fetal development. Because their potential is not total, they are not totipotent and they are not embryos. In fact, if an inner cell mass cell were placed into a woman's uterus, it would not develop into a fetus.

The pluripotent stem cells undergo further specialization into stem cells that are committed to give rise to cells that have a particular function. Examples of this include blood stem cells which give rise to red blood cells, white blood cells, and platelets; and skin stem cells that give rise to the various types of skin cells. These more specialized stem cells are called *multipotent*.

While stem cells are extraordinarily important in early human development, multipotent stem cells are also found in children and adults. For example, consider one of the best understood stem cells, the blood stem cell. Blood stem cells reside in the bone marrow of every child and adult, and in fact, they can be found in very small numbers circulating in the blood stream. Blood stem cells perform the critical role of continually replenishing our supply of blood cells—red blood cells, white blood cells, and platelets—throughout life. A person cannot survive without blood stem cells.

HOW ARE PLURIPOTENT STEM CELLS DERIVED?

At present, human pluripotent cell lines have been developed from two sources[1] with methods previously developed in work with animal models.

(1) In the work done by Dr. [James] Thomson, pluripotent stem cells were isolated directly from the inner cell mass of human embryos at the blastocyst stage. Dr. Thomson received embryos from IVF (In Vitro Fertilization) clinics—these embryos were in excess of the clinical need for infertility treatment. The embryos were made for purposes of reproduction, not research. Informed consent was obtained from the donor couples. Dr. Thomson isolated the inner cell mass and cultured these cells producing a pluripotent stem cell line.

(2) In contrast, Dr. [John] Gearhart isolated pluripotent stem cells from fetal tissue obtained from terminated pregnancies. Informed consent was obtained from the donors after they had independently made the decision to terminate their pregnancy. Dr. Gearhart took cells from the region of the fetus that was destined to develop into the testes or the ovaries. Although the cells developed in Dr. Gearhart's lab and Dr. Thomson's lab were derived from different sources, they appear to be very similar.

The use of somatic cell nuclear transfer (SCNT) may be another way that pluripotent stem cells could be isolated. In studies with animals using SCNT, researchers take a normal animal egg cell and remove the nucleus (cell structure containing the chromosomes). The material left behind in the egg cell contains nutrients and other energy-producing materials that are essential for embryo development. Then, using carefully worked out laboratory conditions, a somatic cell—any cell other than an egg or a sperm cell—is placed next to the egg from which the nucleus had been removed, and the two are fused. The resulting fused cell, and its immediate descendants, are believed to have the full potential to develop into an entire animal, and hence are totipotent. As described these totipotent cells will soon form a blastocyst. Cells from the inner cell mass of this blastocyst could, in theory, be used to develop pluripotent stem cell lines. Indeed, any method by which a human blastocyst is formed could potentially serve as a source of human pluripotent stem cells.

POTENTIAL APPLICATIONS OF PLURIPOTENT STEM CELLS

There are several important reasons why the isolation of human pluripotent stem cells is important to science and to advances in health care. At the most fundamental level, pluripotent stem cells could help us to understand the complex events that occur during human development. A primary goal of this work would be the identification of the factors involved in the cellular decision-making process that results in cell specialization. We know that turning genes on and off is central to this process, but we do not know much about these "decision-making" genes or what turns them on or off. Some of our most serious medical conditions, such as cancer and birth defects, are due to abnormal cell specialization and cell division. A better understanding of normal cell processes will allow us to further delineate the fundamental errors that cause these often deadly illnesses.

Human pluripotent stem cell research could also dramatically change the way we develop drugs and test them for safety. For example, new medications could be initially tested using human cell lines. Cell lines are currently used in this way (for example, cancer cells). Pluripotent stem cells would allow testing in more cell types. This would not replace testing in whole animals and testing in human beings, but it would streamline the process of drug development. Only the drugs that are both safe and appear to have a beneficial effect in cell line testing would graduate to further testing in laboratory animals and human subjects.

Perhaps the most far-reaching potential application of human pluripotent stem cells is the generation of cells and tissue that could be used for so-called cell therapies. Many diseases and disorders result from disruption of cellular function or destruction of tissues of the body. Today, donated organs and tissues are often used to replace ailing or destroyed tissue. Unfortunately, the number of people suffering from these disorders far outstrips the number of organs available for transplantation. Pluripotent stem cells, stimulated to develop into specialized cells, offer the possibility of a renewable source of replacement cells and tissue to treat a myriad of diseases, conditions, and disabilities including Parkinson's and Alzheimer's diseases, spinal cord injury, stroke, burns, heart disease, diabetes, osteoarthritis, and rheumatoid arthritis. There is almost no realm of medicine that might not be touched by this innovation. Some details of two of these examples follow.

- Transplant of healthy heart muscle cells could provide new hope for patients with chronic heart disease whose hearts can no longer pump adequately. The hope is to develop heart muscle cells from human pluripotent stem cells and transplant them into the failing heart muscle in order to augment the function of the failing heart. Preliminary work in mice and other animals has demonstrated that healthy heart muscle cells transplanted into the heart successfully repopulate the heart tissue and work together with the host cells. These experiments show that this type of transplantation is feasible.

- In the many individuals who suffer from Type I diabetes, the production of insulin by specialized pancreatic cells, called islet cells, is disrupted. There is evidence that transplantation of either the entire pancreas or isolated islet cells could mitigate the need for insulin injections. Islet cell lines derived from human pluripotent stem cells could be used for diabetes research and, ultimately, for transplantation.

While this research shows extraordinary promise, there is much to be done before we can realize these innovations. Technological challenges remain before these discoveries can be incorporated into clinical practice. These challenges, though significant, are not insurmountable.

First, we must do the basic research to understand the cellular events that lead to cell specialization in the human, so that we can direct these pluripotent stem cells to become the type(s) of tissue needed for transplantation.

Second, before we can use these cells for transplantation, we must overcome the well-known problem of immune rejection. Because human pluripotent stem cells derived from embryos or fetal tissue would be genetically different from the recipient, future research would need to focus on modifying human pluripotent stem cells to minimize tissue incompatibility or to create tissue banks with the most common tissue-type profiles.

The use of somatic cell nuclear transfer (SCNT) would be another way to overcome the problem of tissue incompatibility for some patients. For example, consider a person with progressive heart failure. Using SCNT, the nucleus of virtually any somatic cell from that patient could be fused with a donor egg cell from which the nucleus had been removed. With proper stimulation the cell would develop into a blastocyst: cells from the inner cell mass could be taken to create a culture of pluripotent cells. These cells could then be stimulated to develop into heart muscle cells. Because the vast majority of genetic information is contained in the nucleus, these cells would be essentially identical genetically to the person with the failing heart. When these heart muscle cells were transplanted back into the patient, there would likely be no rejection and no need to expose the patient to immune-suppressing drugs, which can have toxic effects.

ADULT STEM CELLS

As noted earlier, multipotent stem cells can be found in some types of adult tissue. In fact, stem cells are needed to replenish the supply cells in our body that normally wear out. An example, which was mentioned previously, is the blood stem cell.

Multipotent stem cells have not been found for all types of adult tissue, but discoveries in this area of research are increasing. For example, until recently, it was thought that stem cells were not present in the adult nervous system, but, in recent years, neuronal stem cells have been isolated from the rat and mouse nervous systems. The experience in humans is more limited. In humans, neuronal stem cells have been isolated from fetal tissue and a kind of cell that may be a neuronal stem cell has been isolated from adult brain tissue that was surgically removed for the treatment of epilepsy.

Do Adult Stem Cells Have the Same Potential as Pluripotent Stem Cells?

Until recently, there was little evidence in mammals that multipotent cells such as blood stem cells could change course and produce skin cells, liver cells or any cell other than a blood stem cell or a specific type of blood cell; however, research in animals is leading scientists to question this view.

In animals, it has been shown that some adult stem cells previously thought to be committed to the development of one line of specialized cells are able to develop into other types of specialized cells. For example, recent experiments in mice suggest that when neural stem cells were placed into the bone marrow, they appeared to produce a variety of blood cell types. In addition, studies with rats have indicated that stem cells found in the bone marrow were able to produce liver cells. These exciting findings suggest that even after a stem cell has begun to specialize, the stem cell may, under certain conditions, be more flexible than first thought. At this time, demonstration of the flexibility of adult stem cells has been only observed in animals and limited to a few tissue types.

Why Not Just Pursue Research with Adult Stem Cells?

Research on human adult stem cells suggests that these multipotent cells have great potential for use in both research and in the development of cell therapies. For example, there would be many advantages to using adult stem cells for transplantation. If we could isolate the adult stem cells from a patient, coax them to divide and direct their specialization and then transplant them back into the patient, it is unlikely that such cells would be rejected. The use of adult stem cells for such cell therapies would certainly reduce or even avoid the practice of using stem cells that were derived from human embryos or human fetal tissue, sources that trouble many people on ethical grounds.

While adult stem cells hold real promise, there are some significant limitations to what we may or may not be able to accomplish with them. First of all, stem cells from adults have not been isolated for all tissues of the body. Although many different kinds of multipotent stem cells have been identified, adult stem cells for all cell and tissue types have not yet been found in the adult human. For example, we have not located adult cardiac stem cells or adult pancreatic islet stem cells in humans.

Second, adult stem cells are often present in only minute quantities, are difficult to isolate and purify, and their numbers may decrease with age. For example, brain cells from adults that may be neuronal stem cells have only been obtained by removing a portion of the brain of epileptics, not a trivial procedure.

Any attempt to use stem cells from a patient's own body for treatment would require that stem cells would first have to be isolated from the patient and then grown in culture in sufficient numbers to obtain adequate quantities for treatment. For some acute disorders, there may not be enough time to grow enough cells to use for treatment. In other disorders, caused by a genetic defect, the genetic error would likely be present in the patient's stem cells. Cells from such a patient may not be appropriate for transplantation. There is evidence that stem cells from adults may have not have the same capacity to proliferate as younger cells do. In addition, adult stem cells may contain more DNA abnormalities, caused by exposure to daily living, including sunlight, toxins, and by expected errors made in DNA replication during the course of a lifetime. These potential weaknesses could limit the usefulness of adult stem cells.

Research on the early stages of cell specialization may not be possible with adult stem cells since they appear to be farther along the specialization pathway than pluripotent stem cells. In addition, one adult stem cell line may be able to form several, perhaps 3 or 4, tissue types, but there is no clear evidence that stem cells from adults, human or animal, are pluripotent. In fact, there is no evidence that adult stem cells have the broad potential characteristic of pluripotent stem cells. In order to determine the very best source of many of the specialized cells and tissues of the body for new treatments and even cures, it will be vitally important to study the developmental potential of adult stem cells and compare it to that of pluripotent stem cells.

SUMMARY

Given the enormous promise of stem cells to the development of new therapies for the most devastating diseases, it is important to simultaneously pursue all lines of research. Science and scientists need to search for the very best sources of these cells. When they are identified, regardless of their sources, researchers will use them to pursue the development of new cell therapies.

The development of stem cell lines, both pluripotent and multipotent, that may produce many tissues of the human body is an important scientific breakthrough. It is not too unrealistic to say that this research has the potential to revolutionize the practice of medicine and improve the quality and length of life.

NOTE

1. Michael Shamblott et al., "Derivation of Pluripotent Stem Cells from Cultured Human Primordial Germ Cells," *Proceedings of the National Academy of Sciences USA* 95 (1998): 13726–31; James Thomson et al., "Embryonic Stem Cell Lines Derived from Human Blastocysts," *Science* 282 (1998): 1145–47.

2
The Language Really Matters

Jane Maienschein

On July 31, 2001, the U.S. House of Representatives passed the "Human Cloning Prohibition Act." On August 9, President Bush announced his decision to allow limited research on existing stem cell lines but not on "embryos." In contrast, the U.K. has explicitly authorized "therapeutic cloning." In the blizzard of media attention, much has been said about bioethical, legal, and social implications of these decisions and about whether and how to modify them. The Senate can decline to act on cloning, Congress can decide to overrule the President's Executive order concerning federal funding, and the U.K. can take advantage of U.S. hesitations to ensure their lead in these areas of research. In the meantime, privately funded research and the work that goes on every day in fertility clinics can continue untouched by federal policy.

Throughout these discussions, and despite the NIH's excellent attempts to provide information and education, subtleties of the science have received much less careful consideration.[1] Legislators and reporters struggle with fundamental terms such as "cloning," "pluripotency," "stem cells," and "embryos," as well as whether "adult" are preferable to "embryonic" stem cells as research subjects. They profess to abhor "copying humans" and "killing embryos" but to see scientific research as an extremely important path to progress. Do we all know what we are talking about?

Scientists have largely resisted worrying about definitions not deemed scientifically important, and they have consequently allowed the public to define the central terms of discussion. As a television producer told geneticist Lee Silver, "Clones are not what you think they are." Silver concluded that "Science and scientists would be better served by choosing other words to explain advances in developmental biotechnology to the public."[2] I disagree. Scientists should obviously choose terms that make scientific sense, then should provide clear definitions and be prepared to inform public debate. Without informed scientists' perspectives, we risk making bad policies, in some cases without even realizing that we are doing so because we thought we were talking about something else. In addition, definitions change. They are a matter of convention among a relevant community, and can change either because what we know changes or because the community changes. This is the case for embryos, cloning, and stem cells—for each of which it matters very much what we mean, and for each of which the meanings have changed over times and in different communities. What follows here is a discussion of these terms, in historical context.

EMBRYOS

The first *Encyclopedia Britannica* (1771) defined an "embrio" as "the first rudiments of an animal in the womb, before the several members are distinctly formed; after which period it is denominated a foetus." The word is said to have come from a combination of bryein and em, for swelling inside. This fit a favorite late-eighteenth-century theory that organisms begin essentially as preformed encapsulated forms that swell and grow larger. It was also compatible with the alternative theory of epigenesis, or gradual emergence of form, typically thought to be guided by some vital force or directive agency in the Aristotelian sense of a becoming that actualizes preexisting potential. Both theories regarded the embryo as an incipient organism, requiring time and nutrition to become an adult. It was also generally assumed that at some point in the process of development "quickening" occurs and indicates the beginning of life. The quickening seemed clearly to be the "real" starting point for individual development.

Several steps informed scientific understanding of embryos. In 1827, Karl Ernst von Baer discovered that mammals begin as eggs, as birds so manifestly do, and hence that mammals have continuity with previous generations through that material egg. Despite biologists' best efforts, they could see little about the development of those eggs hidden away inside the mammalian mother until the mid-nineteenth century brought advances in microscopy and histological methods for fixing, staining, and microtoming. During the 1870s, Wilhelm His developed a system for serially sectioning embryos that allowed study of internal structure and organization and led to his pioneering set of human developmental stages.[3] While biologists used the term "embryo" for all stages following fertilization, they saw the stages after formation of the germ layers as the important ones for development of individual organisms, and the earliest stages as largely a matter of cell division and nutrition. The fact that typically the dividing egg remains about the same size up through the blastocyst stage, despite cell divisions, reinforces this interpretation.

By the early twentieth century, biologists had modified both preformationism (now seen in terms of predetermined inherited material passed on from parents to offspring through inherited particles called genes) and epigenesis (the material unfolding of preorganized "organ-forming germ-regions" or the expression of relevant "fates" by cells and germ layers). Determining the balance of predirection and epigenetic emergence was, as Oskar Hertwig put it, *The Biological Problem of To-day*.[4] Biologists prodded, sliced, and diced developing organisms from the germ cell stage through fertilization, morula, blastula, gastrula, and subsequent stages to uncover the processes and patterns of morphogenesis. Along the way, it became much less important what they called an "embryo" or what the domain of "embryology" was thought to be. Usually, the terms referred loosely to all development from fertilization on, but the important stages were thought to be post-gastrulation, which brings the formation of three germ layers.

Whereas leading morphologists including His, Ernst Haeckel, and others emphasized the significance of those germ layers as the starting point for differentiation and individuation, others began to look at a diversity of animals to push back discussions to fertilization and the very first cell division as also important for morphogenesis. Cell lineage, for example, involved tracing lineages of each cell from the unfertilized egg cell itself, though each division until the multiplication of cells made it impossible to follow the details any longer. With meticulous methods for preserving each stage (the egg itself, two cells, four cells, etc.), then observing and drawing or photographing the details, researchers discovered that from the beginning some species have very determinate cleavage patterns suggesting an early differentiation. In contrast, other species allow considerable regulation and demonstrate variability in developmental patterns.

Clearly, morphogenesis is a complex process, with variations across species and even within species depending on the details of an individual case. Throughout this period, morphogenesis, or genesis of form, remained the primary focus of research. The "embryo" receded into the background in importance as a sort of placeholder, while the series of defined stages and the patterns and processes of change gained importance.

One episode still appears in textbooks and class discussions, though we can learn new lessons from the case. Wilhelm Roux and Hans Driesch each experimented on the two-cell stage of an organism: Roux on frogs and Driesch on sea urchins. Roux killed one of the two cells with a hot needle and discovered that the remaining cell develops essentially as it would if it were part of the normal embryo. He concluded that development occurs in a mosaic manner, with individual cells adopting individual fates in the whole organism that follows. Driesch took the two-celled stage of his sea urchins and shook apart the two blastomeres, watching each grow into a larval form as if it were a reorganized whole and not just one half of a mosaic. Driesch concluded that each blastomere at the two and even four cell stage is "totipotent" and capable of considerable regulation in response to changing conditions surrounding the cell.[5]

This difference in their results raised fundamental questions about the nature of differentiation, and led other biologists to take up parallel studies in other organisms for comparison. Some are more mosaic-like, and others more regulative, they discovered. In the period between 1880 and 1910, researchers made tremendous advances in understanding cells and development.[6] Yet they soon reached the limits of those lines of research, and had to set aside what they recognized were exciting questions about regulation and the limits of such totipotency and regulation.

Ross Harrison pursued new methods of exploration by culturing tissues outside the body for the first time; he cultured frog neuroblasts in a lymph medium, and thereby carried out a first step toward current research on precursor and stem cells. Harrison and Hans Spemann also transplanted various parts of donor organisms onto other hosts to discover the relative contributions of each, and this led Spemann to his concept of the "organizer" at the gastrula stage.[7] During this time, deciding what was meant by an "embryo" or "organism" became less important than determining to what extent the stages were defined, predictable, and determined *vs.* regulated in response to environmental conditions.

Around the same time, the ninth *Britannica* (1897) defined "embryology" as "somewhat vaguely applied to the product of generation of any plant or animal which is in process of formation." The term included all stages from fertilized eggs to birth, though a human embryo was seen as becoming a "fetus" at 8 weeks, when it "has assumed the characteristic form and structure of the parent." Textbooks of human embryology outlined successive stages, focusing on appearance of organs and visible features such as limbs and face. Both because of what was possible with existing research approaches and because of what medicine deemed important, later fetal stages remained the primary focus.

We see this for humans in Keibel and Mall's edited classic 1910 *Manual of Human Embryology*. The authors admitted that they knew little of the early stages: "Nothing is known concerning the fertilization of the human ovum, but it may be presumed that it takes place in essentially the same manner as in other animals." "The segmentation stages of the human ovum have not yet been observed." And so on. They were restricted to what they could see with the few selected specimens they had, and these represented later stages, starting with the fourteen day "ovum," which itself revealed little. In later stages they could focus on recognizable organs and the sequence in which they arise. Yet even here, the authors noted,

> It is clear that the normal fully developed organism cannot be produced without a certain regular succession in the development of the organs and a regular interdependence of the individual devel-

> opmental processes, but it is open to question whether this interdependence is the result of the individual and independent development of each organ taking place in such a way that it fits into that of the others or whether it is due to the individual anlagen of an organism mutually influencing one another during the development so as to cause the formation of a normal organism.[8]

These leading human embryologists used the terms "ovum" and "organism," while focusing on the succession of parts and organ. They largely yielded the term "embryo" to non-scientific usage. Biologists instead considered "morphogenesis," "organogenesis," "growth," "differentiation," and such processes. By the 1960s, "embryology" had largely disappeared along with discussion of embryos, giving way to "developmental biology." Encyclopedia definitions also became looser, so that the *Britannica*'s current "embryo" is "the early developmental stage of an animal while it is in the egg or within the uterus of the mother. In humans the term is applied to the unborn child until the end of the seventh week following conception; from the eighth week the unborn child is called a fetus."

This looseness of terminology began to matter a great deal more for political, legal, and social reasons after 1978, when Patrick Steptoe's team demonstrated the astonishing efficacy of in vitro fertilization. They could fertilize human eggs in the laboratory, implant them, and watch mothers bring them to term. They learned to nurture the eggs in glass dishes, select the most viable, freeze and save them, and to manipulate these eggs in many ways. Because of the evident therapeutic benefits to otherwise infertile patients, and perhaps because Steptoe did not announce his efforts until they had been successful, the public accepted this innovation as a medical advance.

The Vatican was not happy, seeing such manipulation of even early developmental stages as morally illicit since Pope Pius IX had in 1869 declared that "ensoulment" and hence life begins at conception. As Karen Dawson noted rather understatedly in a 1990 edited volume on *Embryo Experimentation*, "The debate about embryo research is far from finished and will continue to occur whether governments seek to regulate this rapidly developing area of science."[9] Nonetheless, public opinion in the western world has accepted in vitro medical therapies to allow more people to bear children. Jewish and Muslim scholars, for example, regard life as starting later (forty days is the most commonly accepted time). Typically, the term "embryo" has applied to the stages from fertilization through the eighth week in humans, but there is clearly an important qualitative biological difference between the blastula and gastrula stages.

The U.S. Congress has passed legislation to protect human embryos, the most recent laws prohibiting use of federal funds for "the creation of a human embryo for research purposes or for research in which a human embryo is destroyed." In the Fiscal Year 1998 Appropriations Act, they explicitly defined human embryos to "include any organism, not protected as a human subject under 45 CFR 46 as of the date of the enactment of this Act, that is derived by fertilization, parthenogenesis, cloning, or any other means from one or more human gametes or human diploid cells." Fertilization was clearly thought to be key. Yet it is difficult to discover any clearly reasoned argument leading to this decision, nor any biologically informed discussion of whether subtle distinctions ought to be allowed. Instead, wrapping this decision into omnibus appropriations acts seems to have allowed it to slide through without significant debate. There is room for more careful consideration of what we mean, why, and with what implications.

Up through the blastocyst stage, it is possible to culture the human egg in a glass dish, outside the body. After gastrulation, at least with current technology, it is not. Up until that stage, the blastocyst cells remain largely "plastic" and at least multi or pluripotent (capable of becoming more than one or all cell types). In mice, for example, it has proven possible to combine pairs or small clusters of cells from different organisms and have them join and develop together normally, as a sort of hybrid or chimeric new organism. By gastrulation, this regulative response is no longer possible, at least with present technologies and understanding. The cells have by then begun to differentiate and to become channeled into their respective fate paths.

There is a difference between the blastula and gastrula stages, then, and it is important. Yet we have not developed conventionally accepted terms for these differences. The NIH refers to the entire process as the embryo, while President Bush followed many others in referring to the preimplantation blastula as a "pre-embryo." Clearly, it was important to him to make the distinction. The pre-embryo is different from an embryo: perhaps it is not "alive" in the same way, perhaps it is not an "individual," or conceptually not yet a "potential human" in the same way that an implanted and differentiated postgastrula embryo is. As R. J. Blandon put it in 1971, summarizing Reichert's 1873 studies, the blastocyst is that "structure existing during the period between the end of cleavage and the attachment of the embryo to the uterine wall." It is not itself "the embryo, he declaims, but only the developmental stage of a new the creature, which is the embryo."[10]

Clearly there are biological differences, and much hinges politically on our definitions. Defining a pregastrulation embryo as something different may be the most useful approach. This is certainly the line taken by the British Parliament in 1990 when they agreed to allow some research on the grounds that "such embryos have not progressed to the point that they can be considered as individuals."[11] We could follow Lee Silver's suggestion and seek different terms altogether, but those will not inform the public discussion that relies on familiar and conventionally accepted language. Therefore, we need to work harder to clarify our definitions, to recognize that these will change over time as we learn more, and to develop clearer and scientifically defensible connections between the biological and public language usage. Historians, philosophers, and bioethicists can help with that process.

CLONING

When Ian Wilmut and the Roslin Institute team announced that they had cloned a sheep using a new technique, people around the world scrambled to understand what somatic cell nuclear transfer really means.[12] The image of babies as miniature copies of parents, or as time-delayed twins disturbed many people. After all, the child is not supposed to be the brother of the man—or woman. The collective, if sometimes fascinated, "yuk" suggested that people were not ready for this. And so, as many governments worked to determine whether and what regulations should be imposed, and as Ian Wilmut himself took a lead in speaking out against cloning humans, the U.S. Congress began to generate its own proposed legislation.

Congressman Vernon Ehlers, himself a Ph.D. physicist and one of the few Congressmen with any significant scientific training, proposed a very clean and simple bill that would have outlawed the use of federal funds to clone humans for the purposes of creating a human being. Numerous other bills also appeared, in various stages of development, and a few made their way to committees. I was working as science advisor to our Congressman at the time, and was astonished that some of the early bills were so vaguely worded that they would have outlawed gene cloning or even, in the most extreme case, much of agriculture or even identical twins since one early draft bill proposed to prohibit any genetic replication. Fortunately, in the course of a series of hearings and staff briefings, it became clear that the Congressmen did not know what they were talking about sufficiently to take action. When the Food and Drug Administration declared that they would exercise control over any possible cloning and therefore that no special law was needed, there was a sigh of relief. This held off the perceived need to rush to action. It also helped that Wilmut kept reminding people how difficult the cloning process was, and how many failures occurred before achieving that initial success.

What became very clear very quickly, nonetheless, was the importance of words and of perspective. The concept of cloning was not new. "Cloning" actually first appeared in agriculture, when Herbert John Webber wrote in 1903 of "clons" that were "groups of plants that are propagated by the use of any form of vegetative parts such as bulbs, tubers, cuttings, grafts etc. and are simply parts of the same individual."[13] By the 1920s, cloning covered a variety of asexually reproducing genetic copies, including tree runners, regenerating worms, bacterial divisions, and cell copying in cultures.

In the 1950s, Robert Briggs and Thomas J. King successfully transplanted nuclei from donor *Xenopus laevis* into host frogs. This was cloning through nuclear transplantation, but they could not achieve success with nuclei taken from post-gastrula stages.[14] In the 1960s, John Gurdon significantly expanded the range of techniques and cells, including differentiated frog somatic cells, that could be cloned this way. The work was technically challenging, yielding a low success rate, and attempts to transfer the techniques to mammals failed. The successes showed that there were no natural barriers to such hybrids, but the prospects for human cloning seemed very remote. A brief media blitz gave way to skepticism, while the public probably assumed that such cloning would not happen and relatively few scientists continued to pursue the possibilities.

Textbooks suggested that there was no evidence that it was

impossible to do such cloning in mammals, and even humans. James Watson, John Tooze, and David Jurtz wrote in their *Recombinant DNA: A Short Course* that "In the immediate future there is little likelihood of nuclear transplantation being attempted with any other mammalian species" or even much beyond frogs. "If the efficiency and reproducibility can be improved, the method may, however, find a place in animal breeding. In theory it could be attempted with human eggs and embryonic cells, but for what reason? There is no practical application."[15] That was 1983, and only two years after the first American human in vitro fertilization, so the enormous possibilities in terms of enhancing human fertility were not yet apparent.

The 1950s and 1960s brought cloning of cell lines, with the recognition that there is tremendous advantage in having a large collection of genetically identical cells: both to control experimental variables and to have sufficient material should therapeutic uses be developed. This cell culturing research built on the earlier tissue culture studies that had begun decades earlier in developmental biology. The 1960s brought gene cloning, with the same advantages of genetic duplication. Progressively faster, more accurate, and more efficient laboratory techniques and microbial "factories" have made gene cloning a familiar activity for undergraduate scientists today. The advances have carried us far beyond what Briggs and King could have imagined, but the concepts remain the same. Cloning consistently refers to the full range of types of genetic copying, whatever way that is carried out and on whatever organisms.

Notwithstanding the many efforts to explain that genetic replication does not mean that the expressed offspring cells, tissues, or organisms will be exact copies of the original, the popular perception still sees it that way. As Silver noted, cloning escaped the biologists because of Dolly [the sheep] and the subsequent furor and took on its own public meaning. This does not mean that biologists should give cloning up to the popular press, however. Rather scientists need to continue informing discussions and helping public figures make the distinctions necessary to ground sound policy and legal rulings.

The House bill H.R. 2505, "Human Cloning Prohibition Act of 2001," defines human cloning as "human asexual reproduction, accomplished by introducing nuclear material from one or more human somatic cells into a fertilized or unfertilized oocyte whose nuclear material has been removed or inactivated so as to produce a living organism (at any stage of development) that is genetically virtually identical to an existing or previously existing human organism." Somatic cell "means a diploid cell (having a complete set of chromosomes) obtained or derived from a living or deceased human body at any stage of development." With those definitions, the proposed law states that "It shall be unlawful for any person or entity, public or private, in or affecting interstate commerce, knowingly (1) to perform or attempt to perform human cloning; (2) to participate in an attempt to perform human cloning; or (3) to ship or receive for any purpose an embryo produced by human cloning or any produce derived from such embryo."[16] This bill was passed by the House in part to set the boundaries for decisions about stem cell research and to make it clear that even if such research were allowed, human cloning for the purposes of reproducing humans would be prohibited. To date, the Senate has not enacted any similar bill and seems unlikely to do so without further provocation.

What is important is the boundary constraints that Congress is trying to establish. Far from that initial frightened reaction to the idea of runaway Frankenstein-like scientists copying people in their labs, a more reflective consideration has led to an idea of cloning in terms of somatic cell nuclear transfer. Either the Congress is not aware of the possibility or does not find worrisome that a researcher might transfer a nucleus (or even entire cell) from, for example, a blastocyst into another oocyte from which the nucleus had been removed. This would involve nuclear transfer, but not technically somatic cell nuclear transfer. It could be cloning in the biological sense, since the goal would be genetic copying by the resultant cells. But it is not cloning in the sense of H.R. 2505. Perhaps the Congressmen and their staffs meant to cover this contingency with the definition that somatic cells are derived from the human body during any stage of development. But technically, the blastocyst is not yet a "body," so that still leaves ambiguity.

Cloning clearly includes a range of activities, and that range is likely to expand as we learn more. It is important not to rush to judgment to ban all possible currently foreseeable activities, since we will surely predict imperfectly, nor to allow everything just because we cannot develop crisp definitions. It is also important that scientists not give up the discussions because they are difficult and not central to carrying out the work of each individual scientist. In the long run, it does matter. If legislators are going to make decisions that will restrict some scientific research, then scientists had best be involved in informing those decisions. Historians, philosophers, and bioethicists can help provide perspective, demonstrate what is and is not new, what is and is not good scientific research, and what ethical, legal, and social implications are likely to result.

STEM CELLS

As with embryos and cloning, the term "stem cells" has taken on public meaning. Yet here, perhaps since there is no immediate intuitive sense of the term, that public discussion had been more closely informed by scientific explanation. The primary association is with embryos, since "embryonic stem cells" have moved into the forefront of public debate quite recently.

Also as with embryos and cloning, stem cells and research on them are not new. Yet as with cloning, technical innovations in recent years have excited public imagination. With cloning, Dolly evoked images of sheep-like genetically copied humans. With embryonic stem cell research, the images are of killed embryos on the one hand or of magical laboratory-generated miracle cures on the other. Neither is realistic, nor even close to the whole story.

When Driesch discovered that each of two, or even four separated blastomeres could develop into a whole sea urchin larva, he demonstrated that each cell retains a totipotency, or ability to make the whole. As early twentieth century research placed the material of inheritance in genes, arranged in chromosomes, and passed along through cell and nuclear division to each daughter cell, this raised what Thomas Hunt Morgan called a paradox of development. If all the cells within an organism have the same genetic constitution, what causes them to differentiate? And to what extent is the differentiation determined, or flexible in response to different environmental conditions? This set of questions, essentially a new form of the preformation-epigenesis debates has dominated developmental biology. Exploration of this set of issues lies at the root of stem cell research.

Some cells are able to give rise to an entire, normal, healthy organism. These are totipotent cells, and the fertilized egg is the most obvious. Under some conditions, each of up to the eight cells in humans can do the same, yielding identical octuplets. Again focusing on humans, until the blastocyst stage, there is no reason to believe that any cells are fully differentiated into different cell types. Indeed, the form "blasto" as in blastomere, blastocyst, or blastoderm was used widely by the late nineteenth century for stems, sprouts, or germs, in the sense of able to give rise to form. During the blastocyst stage, it begins to be the case that each cell can no longer give rise to the whole organism by itself. Instead of totipotency, therefore, some of these cells are pluripotent or multipotent, and in later developmental stages following gastrulation, they become differentiated as unipotent, or precursor cells for some particular cell type. Distinctions in these terms are important, and much political ferment hinges on them.

Pluripotent or stem cells are those capable of becoming many, some would say all, of the different cell types of the normal body. This does not mean, however, that any one stem cell can become all the types. Indeed, it cannot. There is no evidence of the capacity and no actual cases in which stem cells give rise to entire organisms. Stem cells, therefore, are not tiny embryos in the rough, nor can they become embryos. They are not "alive" in the sense of giving rise to entire living organisms, though they are capable of reproducing themselves. True pluripotent stem cells can replicate, and can give rise to lines of future stem cells as well as to differentiated cells of a variety of types. Current definitions hold pluripotent stem cells as able to give rise to "most" or "every" cell type, and at least to some types of cells from each of the three germ layers. In practice, however, it is very difficult to prove that any line of stem cells can, in fact, actually give rise to *every* type of cell. Theoretically, however, that is believed possible.

In the 1950s, F. C. Stewart showed that he could generate a whole carrot from single root cells, thereby demonstrating totipotency and raising once again the question just how flexible the most "plastic" cells are. Nuclear transplantation demonstrated a great deal of adaptability in frogs. And, following the discovery in the 1960s that bone marrow transplants allowed patients to produce healthy blood cells to treat anemia and leukemia, the 1960s brought major advances in understanding the multipotent haemotopoietic stem cells. These cells carry out two functions: they differentiate into normal blood cells, and they also maintain and renew the population of stem cells. J. P. Lewis and F. R. Trobaugh argued that they therefore satisfied the definition of stem cells.[17] Their cells were self-renewing as well as differentiating into blood cells, and hence gave rise to two cell types. They were therefore multipotent, and considered true stem cells. From the 1960s until recently, such multipotent bone marrow cell lines have provided our best known example of stem cells. They later proved also to differentiate into other cells useful to the immune system, and therefore to have considerable therapeutic potential as well as demonstrating wider multipotency than originally thought.

Discussions of such stem cells formed a basis for discussion of regeneration. How is it that some organisms, some of the time, and for only some cells, are capable of regenerating injured parts? Crabs can regenerate a claw, but humans cannot regenerate a hand. Worms can regenerate most of their segmented bodies, and snakes regularly regenerate their skins. Why, how, when, under what conditions, and can we use an understanding of such processes to develop medically useful applications, researchers asked. Biologists, who had previously assumed that differentiation works in only one direction as daughter cells cannot go back to a more general state, occasionally questioned that assumption. As Elizabeth Hays wrote in 1974,

> Until better experimental evidence is forthcoming, it no longer is useful, however, to think of differentiation as an irreversible condition, even though we know that somatic cell differentiation is very stable. . . . This is difficult to do because our concepts so overlap that we almost invariably think of differentiation as a restriction of developmental potency, even though we know that cells like the ovum that are truly totipotent actually are differentiated, highly specialized cells.[18]

This was a period of trying to sort out what was meant by stem cells, precursor cells, differentiated and undifferentiated cells, and the extent to which the normal condition could be overruled by experimental manipulation. It was the time when the groundwork for our current discussions began. Unfortunately, it was also a time when some of the definitions remained murky. Whether a stem cell must be multipotent or pluripotent, and what difference that makes were such core questions. What has changed in recent decades is, first, the discovery that stem cells seem to have far wider capabilities than previously thought and, second, that embryonic stem cells appear to have the greatest capacities and there are vast collections of embryos in storage in fertility clinics.

In 1994, the NIH Human Embryo Research Panel recognized the potential in human embryo research, but nonetheless called for moral caution in pursuing this study. The debates might have stalled there, but for the independent publications by James Thomson and John Gearhart in 1998.[19] They showed that stem cell lines could be cultured and could be caused to differentiate into a variety of cell types. As reporter Gretchen Vogel summarized in a special stem cell issue of *Science*, "Conventional wisdom has assumed that once a cell has been programmed to produce a particular tissue, its fate was sealed, and it could not reprogram itself to make another tissue. But in the last year, a number of studies have surprised scientists by showing that stem cells from one tissue, such as brain, could change into another, such as blood." These discoveries served to "signal the unexpected power of stem cells."[20] As leading stem cell researchers hasten to point out, even as they argue for federal funding of their research, many open questions remain: including how much stem cells can be made to do, and whether the most controversial stem cells derived from embryos will prove sufficiently more efficacious to overrule the moral concerns. That we will never learn if we do not ask remains very clear. And that we will not be certain what we have learned unless we are very careful to define our terms and to develop a shared accumulation of knowledge is equally clear.

CONCLUSIONS

Suddenly in the past few years, cloning and stem cell discoveries have heated political discussions about when and what to regulate, and where to draw lines on what procedures scientists should be allowed to pursue. As seen in the preceding discussion, it very much matters what the underlying terms mean. If the public decides to protect embryos, we had better know whether we mean by that fertilized eggs, blastocysts, gastrulas, or only post-implantation stages. Bush and his advisors refer to "pre-embryos," "frozen embryos" stored in fertility clinics, and researchable cell lines from embryos already "destroyed." Bush's August 9 televised speech suggested that he sees cell lines as derived from "destroyed embryos," as already non-living, and therefore as legitimate research subjects, whereas frozen fertilized eggs have a different status, as live embryos. Others would disagree, but in diverse ways. We see considerable reliance on fuzzy definitions to make biologically questionable distinctions that then have tremendous political force.

Clearly it matters what we count as an embryo, what it means to destroy one, and what capacities each developmental stage holds. It matters what we mean by cloning, and how we distinguish embryonic and adult stem cells, and whether they must be pluripotent or multipotent to have therapeutic value. If world leaders and legislators place restrictions on the pursuit of scientific knowledge, the scientific community must work hard to insure the best possible understanding of the science involved. Historians, philosophers, and bioethicists must also work hard and must avoid being seduced by facile distinctions and false hopes growing out of uncorrected misunderstandings and misinterpretations. It is vitally important to define terms clearly and carefully, and for at least a representative set of experts to engage effectively in public discussion not just of the details of their own science but of the words used and the implications of the choices made.

NOTES

1. The National Institutes of Health report, "Stem Cells: Scientific Progress and Future Research Directions," was released in July 2001 in response to a request from Secretary of Health and Human Services Tommy Thompson. The August 11, 2001, *New York Times* includes a photo with a caption, "Karen P. Hughes, counselor to the president, yesterday holding a book that the president used in making his stem cell decision." The NIH Web site contains very valuable and reliable "primers" on stem cell research and other difficult health-related topics.

2. L. Silver, "What Are Clones: They're Not What You Think They Are," *Nature* 412 (July 5, 2001): 21.

3. W. His, *Anatomie Menschlicher Embryonen. I. Embryonen des Ersten Monats* (Leipzig, Ger.: FCW Vogel, 1880).

4. O. Hertwig, *The Biological Problem of Today, Preformation or Epigenesis? A Basis of the Organic Theory*, trans. P. C. Mitchell (London: William Heinemann, 1896). First published as O. Hertwig, *Zeit- und Streitfragen der Biologie* (Jena, Ger.: Fischer, 1894).

5. J. Maienschein, "The Origins of Entwicklungsmechanik," in *A Conceptual History of Modern Embryology*, ed. Scott Gilbert (Baltimore: Johns Hopkins University Press, 1994), pp. 43–61.

6. E. B. Wilson, *The Cell in Development and Inheritance*, 2d ed. 1900, 3d ed. 1925 (New York: Macmillan, 1896).

7. V. Hamburger, *The Heritage of Experimental Embryology* (New York: Oxford University Press, 1988).

8. F. Keibel and F. P. Mall, eds., *Manual of Human Embryology* (Philadelphia and London: Lippincott, 1910), pp. xvi, 18, 980; R. O'Rahilly and F. Müller, *Developmental Stages in Human Embryos* (Washington, D.C.: Carnegie Institution of Washington Publications, 1987), p. 637, includes a revision of Streeter's "Horizons" and a survey of the Carnegie Collection.

9. P. Singer et al., eds., *Embryo Experimentation* (Cambridge: Cambridge University Press, 1990), p. xvi.

10. R. J. Blandon, *The Biology of the Blastocyst* (Chicago: University of Chicago Press, 1971).

11. "The Meaning of Life," *Nature* 412 (July 19, 2001): 255.

12. I. Wilmut et al., "Viable Offspring Derived from Fetal and Adult Mammalian Cells," *Nature* 385 (1997): 810–13.

13. H. J. Webber, "New Horticultural and Agricultural Terms," *Science* 18 (1903): 501–503.

14. R. Briggs and T. J. King, "Transplantation of Living Nuclei from Blastula Cells into Enucleated Frogs' Eggs," *Proceedings of the National Academy of Sciences USA* 38 (1952): 455–63.

15. J. D. Watson, J. Tooze, and D. T. Kurtz, *Recombinant DNA: A Short Course* (New York: Scientific American Books, 1983), pp. 207–208.

16. Congressional Record [online], http://thomas.loc.gov [August 1, 2001].

17. J. P. Lewis and F. E. Trobaugh, "Haematopoietic Stem Cells," *Nature* 204 (1964): 589–90.

18. E. Hays, "Cellular Basis of Regeneration," in *Concepts of Development*, ed. J. Lash and J. R. Wittaker (Sunderland, Mass.: Sinauer Associates, 1974), p. 405.

19. J. A. Thomson et al., "Embryonic Stem Cell Lines Derived from Human Blastocysts," *Science* 282 (1998): 1145–47; J. Gearhart, "New Potential for Human Embryonic Stem Cells," *Science* 282 (1998): 1061–62.

20. G. Vogel, "Can Old Cells Learn New Tricks?" *Science* 287 (2000): 1418.

3
Can Old Cells Learn New Tricks?

Stems cells found in adults show surprising versatility. But it's not clear whether they can match the power of cells from embryos.

Gretchen Vogel

Stem cell biologist Margaret Goodell has never seen her work on muscle and blood development as particularly political, so she was surprised when last month the Coalition of Americans for Research Ethics (CARE), a group that opposes the use of embryos in research, invited her to speak at a congressional briefing in Washington, D.C. She was even more astonished to find herself quoted by conservative columnist George Will a few weeks later.

Goodell gained this sudden notoriety because her work, and that of other teams around the world, just might provide a way around the moral and political quagmire that has engulfed stem cell research to date. Since their discovery in 1998, human embryonic stem cells have been one of the hottest scientific properties around. Because these cells can theoretically be coaxed to differentiate into any type of cell in the body, they open up tantalizing possibilities, such as lab-grown tissues or even replacement organs to treat a

---

variety of human ills, from diabetes to Alzheimer's. Politically, however, human stem cells have been a much tougher sell, as they are derived from embryos or fetuses. Indeed, most research is on hold as policymakers grapple with the ethics of human embryo research.

Enter Goodell, whose work suggests that stem cells derived from adults, in this case, from mouse muscle biopsies, can perform many of the same tricks as embryonic stem (ES) cells can—but without the ethical baggage. Both CARE and George Will seized upon her work as an indication that research on ES cells could remain on hold with no appreciable loss to medicine. "There's a lot less moral ambiguity about the adult stem cells," says bioethicist and CARE member Kevin Fitzgerald of Loyola University Medical Center in Chicago.

But can adult stem cells really fulfill the same potential as embryonic stem cells can? At this stage, the answer is by no means clear. Indeed, scientists caution that it is too early to know if even ES cells will produce the cornucopia of new tissues and organs that some envision. "It is still early days in the human embryonic stem cell world," says stem cell biologist Daniel Marshak of Osiris Therapeutics in Baltimore, which works with adult-derived stem cells.

From a scientific standpoint, adult and embryonic stem cells both have distinct benefits and drawbacks. And harnessing either one will be tough. Although scientists have been working with mouse ES cells for two decades, most work has focused on creating transgenic mice rather than creating lab-grown tissues. Only a handful of groups around the world have discovered how to nudge the cells toward certain desired fates. But that work gained new prominence in late 1998, when two independent teams, led by James Thomson of the University of Wisconsin, Madison, and John Gearhart of The Johns Hopkins University, announced they could grow human stem cells in culture. Suddenly the work in mouse cells could be applied to human cells—in the hope of curing disease.

The beauty of embryonic stem cells lies in their malleability. One of their defining characteristics is their ability to differentiate into any cell type. Indeed, researchers have shown that they can get mouse ES cells to differentiate in lab culture into various tissues, including brain cells and pancreatic cells.

Studies with rodents also indicated that cells derived from ES cells could restore certain missing nerve functions, suggesting the possibility of treating neurological disorders. Last summer, Oliver Brüstle of the University of Bonn Medical Center and Ronald McKay of the U.S. National Institute of Neurological Disorders and Stroke and their colleagues reported that they could coax mouse ES cells to become glial cells, a type of neuronal support cell that produces the neuron-protecting myelin sheath. When the team then injected these cells into the brains of mice that lacked myelin, the transplants produced normal-looking myelin.[1] And in December, a team led by Dennis Choi and John McDonald at Washington University School of Medicine in St. Louis showed that immature nerve cells that were generated from mouse ES cells and transplanted into the damaged spinal cords of rats partially restored the animals' spinal cord function.[2] Although no one has yet published evidence that human ES cells can achieve similar feats, Gearhart says he is working with several groups at Johns Hopkins to test the abilities of his cells in animal models of spinal cord injury and neurodegenerative diseases, including amyotrophic lateral sclerosis and Parkinson's disease.

While Gearhart and his colleagues were grappling with ES cells, Goodell and others were concentrating on adult stem cells. Conventional wisdom had assumed that once a cell had been programmed to produce a particular tissue, its fate was sealed, and it could not reprogram itself to make another tissue. But in the last year, a number of studies have surprised scientists by showing that stem cells from one tissue, such as brain, could change into another, such as blood.[3] Evidence is mounting that the findings are not aberrations but may signal the unexpected power of adult stem cells. For example, Goodell and her colleagues, prompted by the discovery of blood-forming brain cells, found that cells from mouse muscle could repopulate the bloodstream and rescue mice that had received an otherwise lethal dose of radiation.

Bone marrow stem cells may be even more versatile. At the American Society of Hematology meeting in December, hematologist Catherine Verfaillie of the University of Minnesota, Minneapolis, reported that she has isolated cells from the bone marrow of children and adults that seem to have an amazing range of abilities. For instance, Verfaillie and graduate student Morayma Reyes have evidence that the cells can become brain cells and liver cell precursors, plus all three kinds of muscle—heart, skeletal, and smooth. "They are almost like ES cells," she says, in their ability to form different cell types.

These malleable bone marrow cells are rare, Verfaillie admits. She estimates that perhaps one in 10 billion marrow cells has such versatility. And they are only recognizable by their abilities; the team has not yet found a molecular marker that distinguishes the unusually powerful cells from other bone marrow cells. Still, she says, her team has isolated "a handful" of such cells from 80 percent of the bone marrow samples they've taken. Although the versatile cells are more plentiful in children, Verfaillie's team has also found them in donors between forty-five and fifty years old.

Verfaillie's work has not yet been published nor her observations replicated. Even so, many researchers are excited by the work. The cells "look extremely interesting," says hematologist and stem cell researcher Leonard Zon of Children's Hospital in Boston. Stem cell biologist Ihor Lemischka of Princeton University agrees. "I'm very intrigued," he says, although he cautions that data from one lab should not outweigh the decades of research on mouse ES cells.

Besides skirting the ethical dilemmas surrounding research on embryonic and fetal stem cells, adult cells like Verfaillie's might have another advantage: They may be easier to manage. ES cells tend to differentiate spontaneously into all kinds of tissue. When injected under the skin of immune-compromised mice, for example, they grow into teratomas—tumors consisting of numerous cell types, from gut to skin. Before applying the cells in human disease, researchers will have to learn how to get them to produce only the desired cell types. "You don't want teeth or bone in your brain. You don't want muscle in your liver," says stem cell researcher Evan Snyder of Children's Hospital in Boston. In contrast, Verfaillie says her cells are "better behaved." They do not spontaneously differentiate but can be induced to do so by applying appropriate growth factors or other external cues.

Adult stem cells have a drawback, however, in that some seem to lose their ability to divide and differentiate after a time in culture. This short lifespan might make them unsuitable for some medical applications. By contrast, mouse ES cells have a long track record in the lab, says Goodell, and so far it seems that they "are truly infinite in their capacity to divide. There are [mouse] cell lines that have been around for ten years, and there is no evidence that they have lost their 'stem cell-ness' or their potency," she says.

For these and other reasons, many researchers say, adult-derived stem cells are not going to be an exact substitute for embryonic or fetal cells. "There are adult cell types that may have the potential to repopulate a number of different types of tissues," says Goodell. "But that does not mean they are ES cells. Embryonic stem cells have great potential. The last thing we should do is restrict research." Right now, she says, stem cell specialists want to study both adult and embryonic stem cells to find out just what their capabilities might be.

That may be difficult. At the moment, human ES cells are unavailable to most researchers because of proprietary concerns and the uncertain legal status of the cells. Internationally, most research on human ES cells is on hold while legislatures and funding agencies wrestle with the ethical issues. In the United States, the National Institutes of Health is the government agency that would fund the research, and currently, researchers are not allowed to use NIH funds for work with human ES cells. Many European countries, too, are still developing new policies on the use of the cells.[4]

The final version of NIH's guidelines for use of embryonic and fetal stem cells will not appear before early summer, says Lana Skirboll, NIH associate director for science policy. The draft guidelines would allow use of NIH funds for ES cell research as long as the derivation of the cells, by private institutions, met certain ethical standards.[5] But several members of Congress are considering legislation that would overrule the guidelines and block federal funding of ES cell research. At least some of that debate is likely to focus on whether adult stem cells do in fact have the potential to do as much as their embryonic precursors.

NOTES

1. Oliver Brüstle et al., "Embryonic Stem Cell-Derived Glial Precursors: A Source of Myelinating Transplants," *Science* 285 (1999): 754.

2. Ingrid Wickelgren, "Rat Spinal Cord Function Partially Restored," *Science* 286 (1999): 1826.

3. Christopher R. R. Bjornson et al., "Turning Brain into Blood: A Hematopoietic Fate Adopted by Adult Neural Stem Cells in Vivo," *Science* 283 (1999): 534.

4. Noëlle Lenoir, "Viewpoint," *Science* 287 (2000): 1425.

5. Gretchen Vogel, "NIH Sets Rules for Funding Embryonic Stem Cell Research," *Science* 286 (1999): 2050.

4
Adult Stem Cells—A Positive Perspective

Sidney Houff

The use of stem cells to replace adult cells injured by trauma or disease has generated considerable interest. Given their undifferentiated state and proliferate capacity, embryonic stem cells have been the focus of most of the research in cell transplantation. The use of embryonic stem cells has resulted in ethical and legal controversies. These issues resulted in President Bush's decision to limit federal funding to research only existing embryonic stem cells.

Cell transplantation using adult stem cells would circumvent the legal and ethical issues entailed in the use of fetal tissues. But can adult stem cells do the job? I think they can, at least in part. In addition, the knowledge we gain from studying adult stem cells could have a significant impact on embryonic stem cell research.

Do adult stem cells exist and, if so, do they have the same properties found in embryonic stem cells? Adult stem cells have been isolated from a number of organs including bone marrow and brains of rodents, nonhuman primates, and humans. In rodent experiments, differentiation of adult stem cells has resulted in adult cells of the brain as well as other organs. Adult hemapoetic cells have been reported to arise from differentiation of adult stem cells isolated from brain. In humans, donor-derived hepatocytes have been found in the liver of recipients of bone marrow transplants. Further studies have shown that recipients of liver transplants can generate hepatocytes which populated the transplanted liver. These studies suggest the adult animals and humans harbor pluripotential cells capable of generating adult cells that are incorporated into both simple and highly complex cellular architectures. The results of these and other experiments using adult stem cells need confirmation. As with many of the studies with embryonic stem cells, the functional capability of cells derived from adult stem cells needs study.

How do adult stem cells compare to embryonic stem cells? Operationally, stem cells have been defined as being able to differentiate into a wide variety of adult cell types while generating additional stem cells. In many studies, adult stem cells appear not appear to have the ability to generate as many adult cell types or have the proliferative capacity of embryonic stem cells. These findings have suggested that adult stem cells are likely not to be as useful as embryonic stem cells in cell transplantation paradigms. I will propose here that these properties of adult stem cells may actually benefit from our approach to cell transplantation.

Stem cell differentiation into adult cell types occurs through a series of cell divisions associated with increasing phenotypic specificity. If adult stem cells have undergone cell division and differentiation, this could limit the possible adult cell phenotypes they are capable of generating. These partially differentiated stem cells have been called progenitor cells to separate them from true stem cells.

Progenitor cells could offer several advantages in cell transplantation. There are fewer steps to generate the adult cell phenotype needed for tissue repair. Progenitor cells are likely to be more responsive to molecular signals present in adult tissues. The prior commitment of progenitor cells may also facilitate the integration of derived adult cells into the functional architecture of the adult tissue. These properties are likely to be of special importance in transplantation of the central nervous system where anatomical location and cell phenotype are critical for successful neural function. Adult stem cells may offer advantages here as well.

The stem cell "niche" has been emphasized in many studies of the biology of stem cell development. The cellular environment is critically important in stem cell biology. Although the "niche" has yet to be completly defined, molecular signals such as growth factors, cell adhesion molecules, and other molecular signals are known to be important in initiating and supporting stem cell development into adult cell types.

Stem cells, or progenitor cells, present in adult tissues are likely to be more competent to respond to the signals present in the adult stem cell "niche" when compared to embryonic stem cells. It may even be possible that embryonic stem cells or their early derivatives may not have the ability to respond to signals present in adult tissues or require additional signals that are no longer present in adult tissues. Here, the adult stem cell or progenitor is likely to be more successful in repairing tissue damage in the adult. Furthermore, a better understanding of the molecular signals present in the adult stem cell "niche" may be important in understanding cell differentiation beginning with embryonic stem cells.

The study of adult stem cells is likely to lead to many answers important in fulfilling the promise of stem cell biology. They may be able to fulfill the role of tissue repair themselves. The use of adult stem cells, besides avoiding the ethical and legal problems associated with the use of embryonic cells, offer clear advantages in cell transplantation paradigms. The future of stem cell biology, if it is to reach its full potential, will require the study of adult stem cells along with the more conventional approaches using embryonic stem cells.

5
"Parthenotes" Expand the Debate on Stem Cells

Rick Weiss

When is an embryo not an embryo? It's a question fit for a Zen master, but it's one that will have to be answered by federal bureaucrats, researchers, and ethicists as biologists explore the weird world of parthenogenesis.

Parthenogenesis, from the Greek word for "virgin birth," is an unusual mode of procreation in which, to put it plainly, the male might as well find something else to do. Female aphids and turkeys, and certain female reptiles, are among the critters that can reproduce this way. Their eggs can divide on their own as though they had been fertilized by a sperm, then go on to develop into embryos and offspring.

Using chemicals that mimic a sperm's arrival, scientists in recent years have triggered parthenogenesis in the eggs of a few mammals, including rabbits and mice, but the resulting "embryo" has never developed beyond the early fetus stage. Now, with the announcement two weeks ago of the first successful induction of parthenogenesis in human eggs, the Zen riddle arises: Is the resulting doomed entity an embryo, with the same moral standing that some people confer upon conventional and cloned human embryos? Or is it something else—perhaps something that scientists can create and destroy with legal and moral impunity?

This question calls out for an answer—at least on the level of public policy—because the products of human parthenogenesis, which scientists call not embryos but "parthenotes," may prove to be a valuable source of human embryonic stem cells. Stem cells have the capacity to morph into all kinds of cells and tissues, and may someday prove useful in the treatment of degenerative diseases such as diabetes and Parkinson's. There is a roaring political and ethical debate about the use of human embryos as sources of stem cells. Should a similar debate surround work on parthenotes, although there is overwhelming evidence that they lack any potential to become people?

"It forces everybody to get far more nuanced in their thinking about what it is about early life forms that, in their minds, renders them morally significant," said R. Alta Charo, a professor of law and bioethics at the University of Wisconsin in Madison.

The new work was conducted by Michael West and colleagues at Advanced Cell Technology Inc. in Worcester, Massachusetts. In one report, published in an online journal on November 25, the team said it had used chemicals to stimulate human eggs to grow into embryo-like balls of about one hundred cells—the first creation of human parthenotes. (The stimulus was applied before the egg underwent the normal ejection of half its chromosomes, which typically occurs at the time of fertilization to accommodate the sperm's DNA.)

Inexplicably, those parthenotes did not contain any stem cells. But related research, described by West at a scientific meeting on December 2, suggests the goal of obtaining stem cells from human parthenotes is achievable. The team made parthenotes from monkey eggs—the first such success in our near-human relatives—and retrieved from one of those parthenotes what appeared to be stem cells.

Those cells turned into intestine, skeletal muscle, retina, hair follicles, cartilage, bone, and other cell types—even heart cells beating in unison. Some turned into nerve cells that secreted the brain chemical dopamine, the kind of cell that is gradually lost by Parkinson's patients. The idea, West said, is to grow replacement cells and tissues from a patient's own eggs so they are genetically so similar to the woman that they won't be rejected by her immune system.

Men might also someday benefit from parthenote-based therapies, scientists said. Research in animals suggests that male parthenotes can be made by inserting two sperm into an egg whose own DNA has been removed, then stimulating that reconstructed egg to start dividing. Stem cells from the resulting parthenote would be near genetic matches to the male patient who had contributed the sperm.

Technical problems still loom large. No one knows, for example, whether parthenote cells are dangerously inbred because all their genes come from only one parent. "Are they completely normal? Are they prone to metastasize like tumor cells? There's really not enough information," said George Seidel, an embryologist at Colorado State University in Fort Collins. In mammals, both male and female chromosomes are required for normal development.

A practical complication, said Johns Hopkins stem cell researcher John Gearhart, is that viable eggs are difficult to retrieve from postmenopausal women—who are more likely than younger women to be suffering from degenerative diseases.

But Jerry Hall, laboratory director at the Institute for Reproductive Medicine and Genetic Testing in Westwood, California, said his research on mouse parthenotes suggests the cells have promise. "We need to prove they are safe and effective," Hall said. "But I can tell you, they look very normal."

Even if the science goes well, the field will have to navigate ethical and political obstacles. Since 1996, Congress has attached a rider to the Health and Human Services appropriations bill that precludes federal funding of research in which human embryos are destroyed. The rider, attached again to the 2002 bill, which has yet to be signed into law, defines embryos to include parthenotes.

Privately funded research is not affected by that legislation, and some believe that federal funds should be allowed for research on parthenotes that have no potential to grow into people. After all, said Hall, even a skin cell has the potential to become a viable embryo through cloning technology. "If you look at it that way, then a body cell may have more rights or moral basis than a parthenote."

For society to decide what kinds of human entities deserve rights, said Charo, it must consider why rights exist and why society recognizes them. One view, she said, is that rights accrue only if they have some value to the rights holder, leaving embryonic entities with no rights if they lack any potential for life.

Many critics of embryo research don't see it that way, said John Eppig, a researcher at Jackson Laboratory in Bar Harbor, Maine, who sat on a 1995 federal commission that gathered public comments on human embryo research. "The same people who were up in arms about doing research on embryos were up in arms about research on parthenotes," he said. "They correlated this with virgin birth. They correlated it with Christ."

Douglas Johnson of the National Right to Life Committee said he was unconvinced that parthenotes deserve no protection. "With respect to any of these things created with human genetic material," he said, "the burden of proof must rest with those who argue they are not human and that it's okay to kill them."

Part 2
Medical Cures and Promises

Editors' Introduction

Will Superman ever walk again? One of the most tragic stories of 1995 was that of Christopher Reeve, the actor who played Superman in the hit franchise from 1978 to 1983, who fell from his horse and now, from being a truly magnificent specimen of humankind, is paralyzed from the neck down. One of the most heartwarming stories of the last decade has been the way in which Reeve has fought back, not only taking on acting roles but in pushing publicly for research into spinal injuries and related afflictions. (Reeve has since died, in October 2004, but did so while crusading for stem cell research.) Who could fault him for putting his hopes into stem cell research? It is one place where science is making progress for those who need the promise of a cure to keep them going when their bodies have begun to leave them.

We begin this section with some interesting news reports describing how a paralyzed mouse walks again, and how stem cell from umbilical cord blood saved a young man, Keone, suffering from sickle cell anemia. Even if not immediately, it truly now seems that real advances are being made in the quest for cures of spinal injuries—and indeed of many other illnesses and diseases that afflict human beings. From the brain to the spine to the blood and more. All evidenced by Keone and an ambulatory mouse.

To give you some idea of the full scope and potentiality of stem cell research, we turn to the NBAC report to President Clinton, offering an extract detailing some of the possible applications of such investigations. You will see just how broad the range seems to be. Then, taken from *Science*, there is an in-depth look at the way in which Parkinson's disease is being tackled, with information about successes and failures and of the hopes for tomorrow and the barriers still to be surmounted.

We finish up this section with two papers—one that gives advice as to how stem cell therapy going forward should use the lessons from the past, and the other a piece explaining what kind of promises stem cells have kept in the recent past and what it will take in the future to keep these promises.

QUESTIONS TO THINK ABOUT

(1) Stem cells from umbilical cords are very useful as Keone knows. Should saving this blood become a medical imperative or should it remain an elective exercise? What are the benefits and pitfalls of each?
(2) Does the fact that Keone's blood type changed make a difference in his identity or is he the same person?
(3) How much potential should stem cell therapies have before we engage in them?
(4) If the only options for scientific research were animals like cats, dogs, chimps, mice, and other mammals or stem cells, which would you favor and why?
(5) What area contains the biggest concerns you have about the use of stem cells in scientific research?
(6) Does it matter morally that Christopher Reeve and Nancy Reagan wanted advances in stem cell research because it would have benefited them or a loved one like President Reagan?

6
Paralyzed Mouse Walks Again as Scientists Fight Stem Cell Ban

Katty Kay and Mark Henderson

A video showing mice that have been partially cured of paralysis by injections of human stem cells was released last night by American scientists. They are seeking to head off a ban on government funding of similar research.

Researchers at Johns Hopkins University in Baltimore broke with standard scientific practice to screen the tape before details of their research have been formally published, in the hope that it will convince President Bush of the value of stem cell technology.

The U.S. government is considering whether to outlaw all federal funding of studies using stem cells taken from human embryos, which promise to provide new treatments for many conditions, including paralysis and Parkinson's disease.

Opponents argue that the research is immoral as the cells are taken from viable human embryos. President Bush has suspended federal funding of such work and has announced a review of its future. He was urged this week by the Pope to outlaw the practice.

John Gearhart and Douglas Kerr, who led the privately funded research, hope that the tape will have a decisive impact on the debate by showing the potential of the technique. It shows mice paralyzed by motor neuron disease once again able to move their limbs, bear their own weight, and even move around after injections of human embryonic stem cells in their spinal cords.

Dr. Kerr said that the team hopes to start human clinical trials within three years but that a federal funding ban would deal a "potentially fatal blow" to its efforts.

Details of its research were first revealed in November last year, though it has yet to be published in a peer-reviewed journal. In this case, however, the team took the decision to show the tape to Tommy Thompson, the U.S. Health and Human Services secretary, who is conducting a review of stem cell funding for President Bush, and to Pete Domenici, a Republican senator. It is now to be released to the public as well.

Medical research charities said the video would have a major impact. "I wish the president would see this tape," said Michael Manganiello, vice president of the Christopher Reeve Paralysis Foundation, named after the Superman actor who was paralyzed in a riding accident.

"When you see a rat going from dragging his hind legs to walking, it's not that big a leap to look at Christopher Reeve, and think how this might help him," he said.

In the experiment, 120 mice and rats were infected with a virus that caused spinal damage similar to that from motor neuron disease, the debilitating condition that affects Professor Stephen Hawking. The disease is generally incurable and sufferers usually die from it within two to six years.

When fluid containing human embryonic stem cells was infused into the spinal fluid of the paralyzed rodents, every one of the animals regained at least some movement. In previous tests, stem cells have been transplanted directly into the spinal cord. Infusing the fluid is far less invasive and would make eventual treatment in humans much easier.

Dr. Kerr said the limited movement seen was a reflection of the limited research, not of the limits to stem cells themselves.

"I would be a fool to say that the ceiling we have now is the same ceiling we'll see in two years," he said. "We will be smarter and the stem cell research even more developed."

However, the prospect of human trials in three years depends on the outcome of a political and ethical debate over whether the U.S. government will allow federal funding for stem cell research. If President Bush decides not to approve government funds for research, that would set the timetable back ten to twelve years for tests in humans, said Dr. Kerr.

The controversy stems from the fact that human embryos must be destroyed in order to retrieve the stem cells. Mr. Bush is under pressure from conservative Republicans and Roman Catholics not to back the research on moral grounds.

Some top American scientists, who are becoming increasingly frustrated with the funding limitations, have left for Britain where government funding is available. The British government has approved stem cell research on the ground that it could help to cure intractable disease.

The research on rodents at Johns Hopkins took stem cells from five- to nine-week-old human fetuses that had been electively aborted.

7
60 Minutes II, *Holy Grail*

Carol Marin

Stem cells are thought of as the Holy Grail of medicine. One young boy agrees with that. He made medical history because he's been cured of his life-threatening disease. The key to his cure did not come from a human embryo, where all the controversy is, but from something that is routinely tossed in the garbage—an umbilical cord. Umbilical cords were always considered medical waste. Not anymore.

That's why new parents like Pam Dorne and Stephen Ayers of suburban Chicago have decided to save their children's umbilical cord blood. Dorne gave birth last spring to a baby boy, Kyle.

After a baby is born, there is just a fifteen-minute window to retrieve the four to six ounces of blood in the umbilical cord. And in that blood are potentially life saving stem cells that can be saved for future use.

"This is really where, I think, so much of biomedicine is going to be going in the twenty-first century," says Dr. Andrew Yeager of the University of Pittsburgh.

For instance, when stem cells from umbilical cord blood are injected into a person's vein, they migrate to the bone marrow and can create what Dr. Yeager calls a blood factory, replacing diseased blood with healthy blood. According to the National Institutes of Health, stem cells may one day be able repair the body's tissue and muscle and cure everything from spinal cord injuries to Alzheimer's.

"It's not just pie-in-the-sky speculation," says Yeager. "There are studies that would suggest that other organ dysfunction—nerve damage, heart damage, brain-cell damage—might actually be fixed."

It has the potential to make paralyzed patients walk and make Alzheimer's sufferers remember.

That potential is what Dr. Yeager was counting on to cure a young patient named Keone Penn.

Keone suffers from a case of sickle cell, a painful genetic blood disease. He was diagnosed when he was six months old. He was five when his sickle cell caused a stroke.

"All I remember is I woke up and my mama was beside me and there was a basket beside me and a teddy bear," he says. "It was very scary, I mean, whew."

For six years, Keone and his mother, Leslie Penn, were constantly in and out of an Atlanta hospital to receive transfusions to stave off another potentially deadly stroke. By the time Keone was eleven, the transfusions were becoming less effective and he had excruciating pain in his joints and lower back.

"The pain is usually so intense that even morphine, Demerol, those heavy-duty medicines don't really touch it," Leslie Penn says. "All you can really do is pray that he'll just go to sleep."

Keone says he's tough, but at fifteen, he looks much younger. Sickle cell stunted his growth—he's just four feet, nine inches, tall—and restricted what he could do.

"I was impaired from doing a lot of things that normal kids do, like sports or anything or run," he says. "Couldn't play basketball. Because, you know, some people like roughhousing when they play basketball and they can knock you over and push you and that could really hurt me."

The odds were that Keone had, at best, only five years to live. So Yeager decided to take a chance on a new procedure. Never before had stem cells from umbilical cord blood been used to treat sickle cell.

"The goal here is that these stem cells, which are in a relatively high proportion in cord blood—higher than they would be in our own bone marrow and definitely higher than in our own circulating blood—could then be injected and would take hold and again, make more of themselves. And make a whole new blood factory."

Yeager told the family he wasn't sure the procedure would work.

"He just basically said, 'This is just a fifty-fifty chance and it's up to you all if you want to do it, I can't offer you any guarantees,'" recalls Leslie Penn.

Keone Penn remembers how his mother told him: "She came in the room looking very depressed. Pulled the chair up, sat beside me in the bed, and told me everything. And I almost started to cry. But she was very calm about it. She told me everything, said, 'You got—you may—have five years to live,' you know."

Ordinarily, patients with a severe case of sickle cell, like Keone's, would have had a bone-marrow transplant. That's because until recently bone marrow was the only source for stem cells.

But bone marrow transplants can be tricky because there must be a precise match between the person donating the bone marrow and the patient receiving it. In Keone's case, no match could be found.

Stem cells from umbilical cord blood don't need an exact match.

Dr. Yeager and his team found a match that was close enough in a cryogenic tank at the New York Public Blood Bank, which since 1992 has slowly been collecting donations of umbilical cord blood.

Over Christmas vacation of 1998, after intensive chemotherapy to destroy Keone's bad blood, he was injected with the stem cells.

After a few weeks, something extraordinary happened—the stem cells changed his entire blood system from type O to type B.

"That concept there is the one that really blows my mind," says Leslie Penn. "The thought that your whole blood type is changed. The umbilical cord cell's donor, he took on their blood type."

A year later, doctors declared that the sickle cells in Keone's body had disappeared. Today, he is considered cured.

It was umbilical cord stem cells that cured Keone, not stem cells from human embryos. While the use of embryonic stem cells has generated fierce controversy, umbilical cord stem cells have attracted little attention and no political debate. And now it seems, more and more new parents have decided to bank their hopes on the stem cells in their newborn's cord blood.

Moments after Pam Dorne gave birth to baby Kyle, his cord blood was sealed, packed in dry ice, and given to a courier. Within hours, the package was on a plane bound for Tucson, Arizona, where the largest privately run cord blood bank in the country is located.

There, a child's umbilical-cord blood is stored in a cryogenic tank at a temperature of minus 400 degrees Fahrenheit.

Dr. David Harris, laboratory director of the Cord Blood Registry, says it takes only a small vial of cord blood to change a person's entire system.

So far, Cord Blood Registry has collected about thirty thousand samples from families willing to pay a $1,300 flat fee and $95 a year to analyze and privately store their baby's cord blood. The company has taken in over $40 million so far, selling a kind of biological insurance.

"Part of the issue when people bank," says Harris, "is because they have a family history or they work or live in a place where there is a potential for cancer. But part of it is for peace of mind."

According to the American Academy of Pediatrics, that peace of mind isn't worth the money. The academy says the chances a family will ever need to use its frozen cord blood are very small. What they say makes more sense is to donate cord blood to a public bank, the kind where Keone Penn got his stem cells.

That is something Pam Dorne, an obstetrician, says she understands in her professional life. But her own personal choice was a private bank, she says, for one reason.

"If the American Academy of Pediatrics could tell me that none of my children would ever have a problem," she says, "or that if they had a problem, I would be guaranteed that there would be enough donors and somebody would match them, that would be perfectly reasonable. But I don't think anybody has that crystal ball."

What saved Keone Penn's life, Dr. Yeager says, is a public blood bank and the umbilical cord blood from an anonymous donor.

"If they wish to pay, that's absolutely fine," he says of patients. "But to look at a larger, greatest good for greatest number, I would contend that a volunteer donation to a public blood bank would make the most sense."

Meanwhile, Keone, a pioneer, is doing things he's never done before.

"I discovered the other day that I like playing basketball," Keone says. "I never played basketball, 'cause I've always been disabled to play it and to have fun."

Keone, who one day hopes to become a chef, still has some major health problems as a result of infections that occur in most stem-cell transplants. Because of steroids and other medication, he has arthritis, walks with a limp, and will need joint replacement in his hips and knee. But the good news is the sickle cell that was killing him is gone.

"I love stem cells," he says. "I mean they saved my life. If it weren't for them I wouldn't, you never know, I probably wouldn't be here today."

Keone doesn't know where the cord blood came from or who is the owner. He says he would like to know, just so he could say, "Thank You."

8

Human Stem Cell Research and the Potential for Clinical Application

National Bioethics Advisory Commission
Potential Medical Applications

POTENTIAL MEDICAL APPLICATIONS OF HUMAN ES CELL AND EG CELL RESEARCH

Although research into the use of ES and EG cells is still at an early stage, researchers hope to make a contribution to disease treatment in a variety of areas. The ability to elucidate the mechanisms that control cell differentiation is, at the most elemental level, the promise of human ES and EG cell research. This knowledge will facilitate the efficient, directed differentiation of stem cells to specific cell types. The standardized production of large, purified populations of human cells such as cardiomyocytes and neurons, for example, could provide a substantial source of cells for drug discovery and transplantation therapies.[1] Many diseases, such as Parkinson's disease and juvenile-onset diabetes mellitus, result from the death or dysfunction of just one or a few cell types, and the replacement of those cells could offer effective treatment and even cures.

Reprinted from the National Bioethics Advisory Commission (NBAC).

Substantial advances in basic cell and developmental biology are required before it will be possible to direct human ES cells to lineages of human clinical importance. However, progress has already been made in the differentiation of mouse ES cells to neurons, hematopoietic cells, and cardiac muscle.[2] Human ES and EG cells could be put to use in targeting neurodegenerative disorders, diabetes, spinal cord injury, and hematopoietic repopulation, the current treatments for which are either incomplete or create additional complications for those who suffer from them.

USE OF HUMAN ES CELLS AND EG CELLS IN TRANSPLANTATION

One of the major causes of organ transplantation and graft failure is immune rejection, and a likely application of human ES and EG cell research is in the area of transplantation. Although much research remains to be done, ES cells derived through SCNT [somatic cell nuclear transfer] offer the possibility that therapies could be developed from a patient's own cells. In other words, a patient's somatic cells could be fused with an enucleated oocyte and developed to the blastocyst stage, at which point ES cells could be derived for the development of cell-based therapy. This essentially is an autologous transfer. Thus, issues of tissue rejection due to the recognition of foreign proteins by the immune system are avoided entirely. In addition, research to establish xenotransplantation (i.e., interspecies transplantation) as a safe and effective alternative to human organ transplantation is still in its infancy. Alternately, other techniques that would be immunologically compatible for transplantation purposes could be used to generate stem cells, such as

1. banking of multiple cell lines representing a spectrum of major histocompatibility complex (MHC) alleles to serve as a source for MHC matching, and/or
2. creating universal donor lines, in which the MHC genes could be genetically altered so rejection would not occur, an approach that has been tried in the mouse with moderate success.[3]

Autologous transplants would obviate the need for immunosuppressive agents in transplantation as it would decrease a central danger to transplant patients—susceptibility to other diseases. Autologous transplants might address problems ranging from the supply of donor organs to the difficulty of finding matches between donors and recipients. Research on ES cells could lead to cures for diseases that require treatment through transplantation, including autoimmune diseases such as multiple sclerosis, rheumatoid arthritis, and systemic lupus erythematosus. These cells also might hold promise for treating type-I diabetes,[4] which would involve the transplantation of pancreatic islet cells or beta cells produced from autologous ES cells. These cells would enter the pancreas and provide normal insulin production by replacing the failing resident islet cells.

STUDIES OF HUMAN REPRODUCTION AND DEVELOPMENTAL BIOLOGY

Research using human ES and EG cells could offer insights into developmental events that cannot be studied directly in the intact human embryo but that have important consequences in clinical areas, including birth defects, infertility, and pregnancy loss.[5] ES and EG cells provide large quantities of homogeneous material that can be used for biochemical analysis of the patterns of gene expression and the molecular mechanisms of embryonic differentiation.

CANCER THERAPY

Human ES and EG cells may be used to reduce the tissue toxicity brought on by cancer therapy.[6] Already, bone marrow stem cells, representing a more committed stem cell, are used to treat patients after high dose chemotherapy. However, the recovered blood cells appear limited in their ability to recognize abnormal cells, such as cancer cells. It is possible that injections of ES and EG cells would revive the complete immune response to patients undergoing bone marrow transplantation. Current approaches aimed at manipulating the immune system after high-dose chemotherapy so that it recognizes cancer cells specifically have not yet been successful.

DISEASES OF THE NERVOUS SYSTEM

Some believe that in no other area of medicine are the potential benefits of ES and EG cell research greater than in diseases of the nervous system.[7] The most obvious reason is that so many of these diseases result from the loss of nerve cells, and mature nerve cells cannot divide to replace those that are lost. For example, in Parkinson's disease, nerve cells that make the chemical dopamine die; in Alzheimer's disease, it is the cells that make acetylcholine that die; in Huntington's disease the cells that make gamma aminobutyric acid die; in multiple sclerosis, cells that make myelin die; and in amyotrophic lateral sclerosis, the motor nerve cells that activate muscles die. In stroke, brain trauma, spinal cord injury, and cerebral palsy and mental retardation, numerous types of cells are lost with no built-in mechanism for replacing them.

Preliminary results from fetal tissue transplantation trials for Parkinson's disease suggest that supplying new cells to a structure as intricate as the brain can slow or stop disease progression.[8] Yet the difficulty of obtaining enough cells of the right type—that is, dopamine-producing nerve cells—limits the application of this therapy. In 1999, scientists developed methods in animal models to isolate dopamine precursor cells from the dopamine-producing region of the brain and coax them to proliferate for several generations in cell culture. When these cells were implanted into the brains of rodents with experimental Parkinson's disease, the animals showed improvements in their movement control.[9] Scientists also have learned to instruct a stem cell from even a nondopamine region to make dopamine.[10] A large supply of "dopamine-competent" stem cells, such as ES cell lines, could remove the barrier of limited amounts of tissue. (See Exhibit 2-C.)

Another recent development eventually may provide treatments for multiple sclerosis and other diseases that attack the myelin coating of nerves. Scientists have successfully generated glial cells that produce myelin from mouse ES cells.[11] When these ES cell-derived glial cells were transplanted in a rat model of human myelin disease, they were able to interact with host neurons and efficiently myelinate axons in the rat's brain and spinal cord.[12]

Other diseases that might benefit from similar types of approaches include spinal cord injury, epilepsy, stroke, Tay-Sachs disease, and pediatric causes of cerebral palsy and mental retardation. In mice, neural stem cells already have been shown to be effective in replacing cells throughout the brain and in some cases are capable of correcting neurological defects.[13] Human neural stem cells also have recently been isolated and have been shown to be responsive to developmental signals and to be willing to replace neurons when transplanted into mice.[14] These recent discoveries of ways to generate specific types of neural cells from ES cells hold much promise for the treatment of severe neurological disorders that today have no known cure.

Exhibit 2-C: Potential Treatment for Parkinson's Disease

Parkinson's disease is a degenerative brain disease that affects 2 percent of the population over age seventy. Symptoms include slow and stiff movements, problems with balance and walking, and tremor. In more advanced cases, the patient has a fixed, staring expression, walks with a stooped posture and short, shuffling pace, and has difficulty initiating voluntary movements. Falls, difficulty swallowing, incontinence, and dementia may occur in the late stages. Patients often lose the ability to care for themselves and may become bedridden.

The cause of this illness is a deficiency of the neurotransmitter dopamine in specific areas of the brain. Treatment with drugs such as levodopa often is effective in relieving the symptoms. However, as the disease progresses, treatment often becomes more problematic, with irregular responses, difficulty adjusting doses, and the development of side effects such as involuntary writhing movements. Brain surgery with transplantation of human fetal tissue has shown promise as therapy.

Stem cell transplantation also may be a promising therapy for Parkinson's disease. The injection of stem cells that can differentiate into brain cells may offer a means of replenishing neurons that are capable of synthesizing the deficient neurotransmitter. It is possible that stem cell transplantation may be simpler and more readily available than fetal tissue transplantation.

DISEASES OF THE BONE AND CARTILAGE

Because ES and EG cells constitute a relatively self-renewing population of cells, they can be cultured to generate greater numbers of bone or cartilage cells than could be obtained from a tissue sample. If a self-renewing, but controlled, population of stem cells can be established in a transplant recipient, it could effect long-term correction of many diseases and degenerative conditions in which bone or cartilage cells are deficient in numbers or defective in function. This could be done either by transplanting ES and EG cells to a recipient or by genetically modifying a person's own stem cells and returning them to the marrow. Such approaches hold promise for the treatment of genetic disorders of bone and cartilage, such as osteogenesis imperfecta and the various chondrodysplasias. In a somewhat different potential application, stem cells perhaps could be stimulated in culture to develop into either bone- or cartilage-producing cells. These cells could then be introduced into the damaged areas of joint cartilage in cases of osteoarthritis or into large gaps in bone that can arise from fractures or surgery. This sort of repair would have a number of advantages over the current practice of tissue grafting.[15]

BLOOD DISORDERS

The globin proteins are essential for transport of oxygen in the blood, with different globins expressed at different developmental stages. The epsilon globin gene is expressed only in embryonic red blood cells. When this gene—which is not normally expressed in the adult—is artificially turned on in sickle cell patients, it blocks the sickling of the cells that contain sickle cell hemoglobin. Research involving ES cells could help answer questions about how to turn on the epsilon globin gene in adult blood cells and thereby halt the disease process. Stem cell research also may help produce transplantable cells that would not contain the sickle cell mutation.

TOXICITY AND DRUG TESTING

Human stem cell research offers promise for use in testing the beneficial and toxic effects of biologicals, chemicals, and drugs in the most relevant species for clinical validity—humans. Such studies could lead to fewer, less costly, and better designed human clinical trials yielding more specific diagnostic procedures and more effective systemic therapies. Beyond the drug development screening of pharmacological agents for toxicity and/or efficacy, human stem cell research could define new research approaches for clarifying the complex association of environmental agents with human disease processes.[16] It also makes possible a new means of conducting detailed investigations of the underlying mechanisms of the effects of environmental toxins or mixtures of toxins, including their subtle effects on the developing embryonic and fetal development tissue systems.

TRANSPLANTABLE ORGANS

Several researchers are investigating ways to isolate AS cells and create transplantable organs that may be used to treat a multitude of diseases that do not rely upon the use of embryonic or fetal tissue. Moreover, if it is found to be possible to differentiate ES cells into specific cell types, such stem cells could be an important source of cells for organ growth. For example, recent developments in animals have shown that it may be possible to create entire transplantable organs from a tissue base in a manner that would overcome such problems as the limited supply of organs and tissue rejection. Such a development—producing this tissue base by directing the growth of human embryonic cells—could be a major breakthrough in the field of whole organ transplantation.

For example, using tissue engineering methods, researchers have successfully grown bladders in the laboratory, implanted them into dogs, and shown them to be functional.[17] To create the bladders, small biopsies of tissue were taken from dog bladders. The biopsied tissue was then teased apart to isolate the urothelial tissue and muscle tissue, which were then grown separately in culture.[18] The tissue was then applied to a mold of biodegradable material with the urothelial tissue on the inside and the muscle tissue on the outside. The new organs were transplanted within five weeks.[19]

Dogs that received the tissue-engineered organs regained 95 percent of their original bladder capacity, were continent, and voided normally. When the new organs were examined eleven months later, they were completely covered with urothelial and muscle tissue and had both nerve and blood vessel growth. Dogs that did not undergo reconstructive procedures or only received implants of the biodegradable molds did not regain normal bladder function.[20] This accomplishment marks the first time a mammalian organ has been grown in a laboratory. The ability to create new organs by seeding molds with cells of specific tissue types would be extremely useful in treating children with congenital malformations of organs and people who have lost organs due to trauma or disease.[21]

SUMMARY

Currently, human ES cells can be derived from the inner cell mass of a blastocyst (those cells within the conceptus that form the embryo proper), and EG cells can be derived from the primordial germ cells of fetuses. These cells, present in the earliest stages of embryo and fetal development, can generate all of the human cell types and are capable, at least for some time, of self-renewal. A relatively renewable tissue culture source of human cells that can be used to generate a wide variety of cell types would have broad applications in basic research, transplantation, and other important therapies, and a major step in realizing this goal was taken in 1998 with the demonstration that human ES and EG cells can be grown in culture. The clinical potential for these stem cells is vast—they will be important for in vitro studies of normal human embryogenesis, human gene discovery, and drug and teratogen testing and as a renewable source of cells for tissue transplantation, cell replacement, and gene therapies.

NOTES

1. J. A. Thomson et al., "Embryonic Stem Cell Lines Derived from Human Blastocysts," *Science* 282 (1998): 1145–47.

2. O. Brustle et al., "*In vitro*-generated Neural Precursors Participate in Mammalian Brain Development," *Proceedings of the National Academy of Sciences USA* 94 (1997): 14809–14.

3. National Institute of Allergy and Infectious Diseases (NIAID), "What Would You Hope to Achieve from Stem Cell Research?" Response to Senator Specter's Inquiry, March 25, 1999.

4. Senate Appropriations Subcommittee on Labor, Health, and Human Services, January 12, 1999, D. Melton, *Testimony of the Juvenile Diabetes Foundaton*; Senate Appropriations Subcommittee on Labor, Health, and Human Services, December 2, 1998, H. Varmus, *Testimony*.

5. Thomson et al., "Embryonic Stem Cell Lines."

6. National Cancer Institute (NCI), "What Would You Hope to Achieve from Stem Cell Research?" Response to Senator Specter's Inquiry, March 25, 1999.

7. J. D. Gearhart, "New Potential for Human Embryonic Stem Cells," *Science* 282 (1998): 1061–61; Varmus, *Testimony*.

8. C. R. Freed et al., "Double-Blind Controlled Trial of Human Embryonic Dopamine Cell Transplants in Advanced Parkinson's Disease: Study Design, Surgical Strategy, Patient Demographics, and Pathological Outcome" (presented to the American Academy of Neurology, April 21, 1999).

9. National Institute of Neurological Disorders and Stroke (NINDS), "What Would You Hope to Achieve from Stem Cell Research?" Response to Senator Specter's Inquiry, March 25, 1999.

10. J. Wagner et al., "Induction of a Midbrain Dopaminergic Phenotype in *Nurr1*-Overexpressing Neural Stem Cells by Type 1 Astrocytes," *Nature Biotechnology* 17 (1999): 653–59.

11. O. Brustle et al., "Embryonic Stem Cell–Derived Glial Precursors: A Source of Myelinating Transplants," *Science* 285 (1999): 754–56.

12. Ibid.

13. H. D. Lacorazza et al., "Expression of Human b-hexosaminidase a Subunit Gene (the Gene Defect of Tay-Sachs Disease) in Mouse Brains upon Engraftment of Transduced Progenitor Cells," *Nature Medicine* 2 (1996): 424–29; C. M. Rosario et al., "Differentiation of Engrafted Multipotent Neural Progenitors towards Replacement of Missing Granule Neurons in Meander Tail Cerebellum May Help Determine the Locus of Mutant Gene Action," *Development* 124 (1997): 4213–24; E. Y. Snyder et al., "Multipotent Neural Precursors Can Differentiate toward Replacement of Neurons Undergoing Targeted Apoptotic Degeneration in Adult Mouse Neocortex," *Proceedings of the National Academy of Sciences USA* 94 (1997): 11663–68; E. Y. Snyder, R. M. Taylor, and J. H. Wolfe, "Neural Progenitor Cell Engraftment Corrects Lysosomal Storage Throughout the MPS VII Mouse Brain," *Nature* 374 (1995): 367–70; B. D. Yandava, L. L. Billinghurst, and E. Y. Snyder, "'Global' Cell Replacement Is Feasible via Neural Stem Cell Transplantation: Evidence from the Dysmyelinated *Shiverer* Mouse Brain," *Proceedings of the National Academy of Sciences USA* 96 (1999): 7029–34.

14. J. D. Flax et al., "Engraftable Human Neural Stem Cells Respond to

Developmental Cues, Replace Neurons, and Express Foreign Genes," *Nature Biotechnology* 16 (1998): 1033–39.

15. National Institute of Arthritis and Musculoskeletal and Skin Diseases (NIAMS), "What Would You Hope to Achieve from Stem Cell Research?" Response to Senator Specter's Inquiry, March 25, 1999.

16. National Institute of Environmental Health Sciences (NIEHS), "What Would You Hope to Achieve from Stem Cell Research?" Response to Senator Specter's Inquiry, March 25, 1999.

17. F. Oberpenning et al., "De Novo Reconstruction of a Functional Mammalian Urinary Bladder by Tissue Engineering," *Nature Biotechnology* 17 (1999): 149–55.

18. J. H. Tanne, "Researchers Implant Tissue-Engineered Bladders," *British Medical Journal* 318 (1999): 350B.

19. Ibid.

20. Oberpenning et al., "De Novo Reconstruction."

21. Tanne, "Researchers Implant Tissue-Engineered Bladders."

9

Fetal Neuron Grafts Pave the Way for Stem Cell Therapies

A decade of experimental treatments using fetal neurons to replace brain cells that die in Parkinson's disease can provide lessons for planning stem cell therapies.

Marcia Barinaga

Swedish neuroscientist Anders Björklund and his colleagues may have caught a glimpse of what the future holds for the treatment of failing organs. For more than ten years, Björklund has been part of a team at Lund University in Sweden that has been grafting neurons from aborted fetuses into the brains of patients with Parkinson's disease. In many cases, the transplanted cells have dramatically relieved the patients' symptoms, which include slowness of movement and rigidity. That is just the kind of therapy that stem cell researchers hope to make routine for treating all sorts of degenerative diseases, if they can coax the cells to develop into limitless supplies of specific cell types that can be used to repair or replace damaged organs.

Although the current Parkinson's treatment uses fetal cells that have already developed into a particular type of neuron, the promising results represent a "proof of principle that cell replacement actually works," says Björklund. The results have given researchers increased confidence that, if they can manipulate stem cells to

Reprinted with permission from *Science* 287, no. 5457 (2000): 1421–22.

develop into the kind of neuron the Lund group and others are using—a big challenge—the new cells would take over the work of damaged cells in the brains of Parkinson's patients. If so, Parkinson's treatment could be among the first applications of stem cell therapy.

The successes have also increased the urgency of developing stem cell treatments, because despite their promise, there are many reasons that fetal cells will never be widely used to treat Parkinson's disease. The reasons range from ethical concerns, such as the protests by antiabortionists that led the governor of Nebraska to urge that research involving fetal tissue be shut down at the University of Nebraska,[1] to the fact that there will never be enough fetal tissue to treat all the people who need it. Parkinson's disease afflicts 1 million people in the United States alone.

Researchers are now looking closely at the results from fetal cell transplants for lessons that will help guide future work with stem cells. There are still many hurdles to overcome, but this first round of cell replacement in the brain sets a "gold standard" that stem cells must meet if they are to become the basis for new Parkinson's treatments, says neuroscientist and stem cell researcher Evan Snyder of Harvard Medical School in Boston.

Parkinson's disease is a logical candidate for cell replacement therapy, in part because conventional treatments have had limited success. The disease is caused by the death, for unknown reasons, of a particular group of brain neurons that produce dopamine, one of the chemicals that transmit signals between nerve cells. Afflicted people lose the ability to control their movements, ultimately becoming rigid. Treatment with levodopa (L-dopa), a drug that is converted to dopamine by the brain, alleviates these symptoms, but as the neurons continue to die, L-dopa's effectiveness wanes. Researchers first tried replacing the dopamine-producing cells by grafting into the affected region cells from the adrenal medulla gland. These cells are not neurons, but they make dopamine and can be coaxed to become neuronlike. The treatment reversed Parkinson's symptoms in rats, but produced little lasting improvement in human patients, probably because the cells died or stopped making dopamine, says John Sladek, chair of neuroscience at the Chicago Medical School.

Researchers have had better luck grafting immature neurons

taken from aborted human fetuses. Dozens of patients who have received these experimental dopamine-neuron grafts over the past ten years have had up to a 50 percent reduction in their symptoms. And the effects appear to last. Using positron emission tomography to image the brain, Olle Lindvall of Lund University and a team of colleagues in Lund and at Hammersmith Hospital in London reported in the December issue of *Nature Neuroscience* that, in one patient, the transplanted neurons are still alive and making dopamine ten years after the surgery. That's encouraging, says neurotransplant researcher Ole Isacson of Harvard Medical School: Whatever killed the brain's own dopamine-producing neurons doesn't seem to have killed the transplanted cells.

Still, fetal cell transplants are plagued by problems that can never be overcome. Aside from ethical concerns about scavenging neurons from aborted fetuses, there are practical issues. It takes six fetuses to provide enough material to treat one Parkinson's patient, in part because as many as 90 to 95 percent of the neurons die shortly after they are grafted. Indeed, Lund's Björklund says, the cell supply is so limited that researchers have not even been able to test some possible avenues for fetal cell transplants. The neurons that die in Parkinson's originate in a brain region called the substantia nigra and send their long axons to several other areas, where they release dopamine. So far, researchers have put the cell grafts into only one of these areas, the putamen—and even there, they have not yet transplanted enough neurons to restore normal dopamine levels in most cases.

Even if researchers can develop techniques that diminish the fetal cell die-off, there will never be enough fetuses available to make this an "everyday procedure," says Sladek. What's more, the brain material recovered from aborted fetuses "comes out in a form that makes it difficult to . . . standardize" in terms of quality and purity, Björklund says. This is likely why some patients do far better than others—uncertainty that would be unacceptable in a standard medical treatment.

Consequently, researchers are pinning their hopes on cultured stem cells. They would eliminate a continuing dependence on aborted fetuses, although the ethical concerns won't be completely laid to rest unless researchers can use stem cells obtained from adults rather than embryos.[2] And the supply of cultured cells could be unlimited, allowing tests of grafts into the putamen and possibly into other brain areas as well. The cell treatment, moreover, could be standardized and controlled to assure a more predictable outcome. "The ability to grow the cells of interest will make this a routine technology," predicts neuroscientist Ron McKay, whose team is working on ways to culture neural stem cells at the National Institute of Neurological Disorders and Stroke.

But to make this brave new world of cell replacement technology a reality, researchers must first learn how to keep stem cells dividing for many generations in culture and then be able to trigger them to differentiate into the type of neuron they want. Stem cells presumably have the ability to differentiate into any of the several different types of dopamine neurons the brain contains, but it may be crucial to use the specific type of dopamine neurons that die in Parkinson's. Researchers doing the fetal cell transplants specifically select these neurons—known as nigral neurons because they originate in the substantia nigra—when they harvest neurons for grafting from fetal brains. Nigral neurons are "genetically programmed and designed to be a dopamine neuron in the appropriate brain circuit," Sladek says.

Among other things, nigral neurons may respond better to local conditions, producing just enough dopamine. Experience with L-dopa treatment has shown that too much of the neurotransmitter can be just as problematic as too little, causing uncontrollable, jerky movements in patients. Researchers worry that stem cells coaxed to develop into dopamine neurons may become one of the nonnigral types and will not regulate their dopamine output in the appropriate ways. "It is like putting the right alternator into your car," Sladek says. "If you put in one designed for another model of car, it may not work as well."

Getting stem cells to differentiate into the right type of neuron may be only part of the problem. Neurons in their natural environment are surrounded by support cells called "glia," which nurture the neurons and even modulate their activity, and optimal cell transplants may require replacing not only dopamine neurons but also the glia that normally surround them, Harvard's Snyder suggests.

But some researchers believe that the brain itself may be able to overcome the hurdles of producing both the proper neurons and the support cells they need. Snyder's lab has shown in animal experiments that stem cells put into the brain can be influenced by the brain environment to differentiate into both neurons and support cells. He envisions someday putting stem cells into Parkinson's brains and letting the brain tell them which cell types to become.

Even if it turns out not to be quite that simple, Parkinson's poses a much less daunting challenge for cell replacement therapy than do other neurological disorders. "The dopaminergic system is a fairly easy system to work with compared to sensorimotor or visual systems or spinal cord," says Isacson. That, he says, is because the nigral neurons lost in Parkinson's disease have a diffuse and relatively nonspecific network of connections in the brain areas they link up to, rather than the very intricate and precise connections made by neurons in many other parts of the nervous system.

The treatment of most other brain disorders would likely require coaxing new neurons to make very precise connections, a task that no one is sure how to achieve. But in the case of Parkinson's disease, simply getting neurons to release dopamine in the correct general area helps patients. Because of that difference, Isacson predicts that "it will take some time to get other diseases to benefit from all these discoveries." Nevertheless, a successful Parkinson's treatment based on stem cells would still be a dramatic achievement. "You would help a huge number of patients," he says, "as many as the surgeons could do."

NOTES

1. E. Marshall, "Antiabortion Groups Target Neuroscience Study at Nebraska," *Science* 287 (2000): 202.

2. G. Vogel, "Can Old Cells Learn New Tricks?" *Science* 287 (February 25, 2000): 1418.

10
Promises of Stem Cells Kept

Leigh Shoemaker

When one refers to the "promises" of stem cells, it's easy to imagine the sorts of things to which the speaker is referring: promises of an almost miraculous set of treatments for a plethora of currently uncurable ailments and maladies such as diabetes, stroke, Parkinson's, terminal cancers, and the like. But what sort of "promises" are these? Clearly, the stem cells themselves are not making promises to their human bearers; they have no intention and no ability to act upon or to break a promise. If anything, the "promise" of stem cells is something more akin to "potential." Given the current state of research into the potential of stem cells, we actually know quite little about the full potential of stem cells. Rather, we have whispers and intimations of the sorts of amazing medical breakthroughs that stem cells are likely to allow. Unfortunately, at this time we have very little empirical evidence of the fulfillment of this potential in human subjects.

If anyone is making promises regarding stem cells, it is the scientists and physicians who are researching their potential, engaging in clinical trials to determine whether and to what degree that potential can be actualized, and adding their voices to those of advocates for federally funded research into the potential of embryonic stem cells. It is this latter activity that undergirds the majority of the public discussion of the promises of stem cells, and it also frames the discussion regarding whether and how the promises of stem cells have been kept.

WHAT ARE STEM CELLS?

Before entering into a discussion of the various promises of stem cell research, it is important to clarify the terms of the discussion. When one hears of "stem cell research" in popular media, it is typically within the context of controversy regarding the practice of harvesting and studying embryonic stem cells. While these cells are widely recognized to hold the most potential for stem cell therapy, they are not the only type of stem cells, nor are they the only stem cells that hold promise for medical research and application.

All stem cells share some fundamental and unique properties: they are capable of dividing and renewing themselves for long periods; they are unspecialized, or "undifferentiated"; and they can give rise to specialized, or "differentiated," cell types.[1] Both of the major types of stem cells—embryonic and adult—hold promise for medical therapy; however, embryonic stem cells are thought to hold the most promise.

Embryonic stem cells are "pluripotent"—that is, they have the potential to give rise to almost any cell type in the body. This is why the weight of the promise of stem cell research lies primarily with them. Embryonic stem cells must be harvested from human embryos (typically four or five days old), a procedure which destroys the embryo, thereby generating controversy among members of the medical, political, and religious communities, as well as the public at large. Since 2001, federally funded research on embryonic stem cells has been restricted to existing lines of stem cells isolated by researchers late in the twentieth century. Unfortunately, the quality of these lines has deteriorated greatly over time, and their value for research is questionable at best. In spring of 2005, a bill passed the U.S. House of Representatives that would expand the number of stem cell lines eligible for federally funded research by allowing excess embryos produced at in vitro fertilization clinics to be donated for research by prospective parents. As of December 2005, this bill has yet to reach the Senate. However, even if it were to pass the Senate, it faces the promise of a veto from President George W. Bush.

Adult stem cells, on the other hand, are not pluripotent. They are "multipotent"—they can differentiate into only a limited number of cell types. Typically, adult stem cells can only generate the cell types of the tissue in which they reside. While these stem cells may well prove to be more manipulable than once thought, their potential is quite limited compared to that of pluripotent stem cells. They "have not yet been isolated from all cell and tissue types, and they have not been shown to be capable of developing into all of the different cell and tissue types of the body. Furthermore, adult stem cells are difficult to obtain . . . difficult to isolate and purify . . . their numbers appear to decrease with age . . . [they] may have more DNA damage, and they appear to have a shorter lifespan than pluripotent stem cells."[2]

The potential of pluripotent stem cells, combined with the current ban on federal funding for research using new lines of embryonic stem cells, has motivated scientists to explore various workarounds to the controversy surrounding the harvesting of pluripotent stem cells from blastocysts. For instance, much effort is currently being directed at identifying greater plasticity among adult stem cells. Along these lines, researchers have recently determined that stem cells found in adult bone marrow may have a plasticity similar to that of embryonic stem cells, a discovery which holds great potential and promise for ongoing research. Taking a different angle, researchers at the Stanford University Institute for Stem Cell Biology and Regenerative Medicine propose the practice of transferring a nucleus from an adult cell into an existing stem cell line as a way to avoid destroying a blastocyst and the attendant controversy that surrounds that destruction. However, this process has not yet been successful in any organism.[3]

THE PROMISES

Clearly, most stem cell research is conducted with an eye to developing clinical applications and therapies that may one day be used to treat a wide range of diseases, conditions, and disabilities including Parkinson's, spinal cord injury, stroke, heart disease, diabetes, osteoarthritis, rheumatoid arthritis, autoimmune diseases, and burns. It is also motivated by a desire to expand the knowledge base regarding both human development and disease in general. It should come as no surprise that due to the limits on federal funding for research using embryonic stem cells, most of the stem cell therapies currently in use are based on research conducted with adult stem cells. However, adult stem cell research has been pursued for more than thirty-five years and has yielded only a few successful applications.

Adult bone marrow (hematopoietic) stem cells are currently used to treat leukemia, lymphoma, and several inherited blood disorders. More recent uses include treatment for diabetes and kidney cancer; however, these latter applications have only been conducted with a very limited number of patients.

Research into the promise of adult stem cells continues. In 2000, Canadian researchers reported success in transplanting pancreatic islet cells in patients with "Type I diabetes and a history of severe hypoglycemia and metabolic instability."[4] All seven patients "quickly attained sustained insulin independence." However, development of this therapy is limited by a shortage of human donor tissue. In August 2004, Canadian researchers located immature cells in the pancreas of adult mice that were able to develop into cells that produce insulin.[5] In December 2005, researchers at the University of Louisville reported finding cells in adult bone marrow that appear to mimic embryonic stem cells. If these results of the Canadian research can be duplicated in human subjects, or if the preliminary success in growing and differentiating the bone marrow cells can be sustained, the promise of adult stem cells will be greatly expanded.

In a July 2004 testimony before the Senate Commerce Committee Subcommittee on Science, Technology, and Space, Robert Goldstein, chief scientific officer of the Juvenile Diabetes Research Foundation (JDRF), spoke of the promise of embryonic stem cell research in the area of diabetes. He cited recent studies demonstrating the ability of embryonic stem cells to grow insulin-producing cells in a laboratory environment. The promise is that large amounts of these cells could be grown and transplanted into patients. Goldstein noted that adult stem cells do not show the same promise as embryonic stem cells for treating diabetes.

Research into the promise of embryonic stem cells is proceeding at a more measured pace. However, the political ban on funding does not lessen the medical promise of these cells. What is limited is the ability of humans to isolate, unlock, and act upon those promises.

THE ACT OF PROMISING

As previously mentioned, stem cells are not capable of making a promise (or promises) per se. When the term "promise" is used in the context of stem cells, it is used colloquially to refer to their *potential*. This potential, particularly in the case of pluripotent stem cells, is widely acknowledged to be great. Research conducted on stem cells holds great potential for the study of human disease as well as for therapeutic and clinical applications. The latter is the more pressing need, because there are many patients currently afflicted with painful and irreversible disorders and diseases who have a direct interest in the great potential of stem cells to ease—or even to cure—their discomfort.

Given the dire situations of many of these patients—individuals with Type I diabetes, Parkinson's and other neurological disorders, terminal cancers, spinal cord injuries, severe burn trauma, and the like—and given the potential for healing that pluripotent stem cells offers to these people and others like them, does the *potential* of stem cells imply a *promise*?

It is the case that physicians and other medical personnel and caregivers make implicit promises to their patients simply by assuming their respective mantles. It is common to speak of a physician's "duty" to her patients; the Hippocratic Oath (an example of an explicit promise made by physicians) states that a physician "will apply, for the benefit of the sick, all measures which are required. . . ." Patients expect physicians to take care of their bodies, to provide them with the best treatments available to aid in healing—and where treatments are not available, to seek additional therapies that will benefit them and others when in need. Do the implicit and explicit promises made by caregivers also extend to stem cells, particularly those pluripotent stem cells with the most potential to ease human suffering? With regard to stem cells, are physicians and other scientists keeping their promises?

Arguably, they are. Research into the "promise" of stem cells, both adult and embryonic, is currently being pursued despite political hurdles. Physicians are working to gain insight into the potential of stem cells, both adult and embryonic, for medical therapies and treatments. Whether this work is proceeding as swiftly as it could, or whether valuable scientific energies are being misdirected due to political or religious controversy are different (albeit related) questions. The scientific process of experimentation, repetition, discussion, review, and approval is a notoriously slow one—and in the case of promising new medical therapies with the potential to ease great suffering, this process can be and often is a frustrating one, particularly when it is coupled with questions of morality, religion, politics, and community standards. But important work is being done—regardless of whether current generations will benefit from its conclusions—and this work must be both encouraged and allowed to continue. And that is the promise of stem cells kept.

NOTES

1. "Stem Cell Information: The Official National Institutes of Health Resource for Stem Cell Research" [online], http://stemcells.nih.gov.
2. "NIH Fact Sheet on Human Pluripotent Stem Cell Research Guidelines" [online], http://stemcells.nih.gov/news/newsArchives/stemfact sheet.asp.
3. Stanford School of Medicine, Stanford University, "Q&A on the Institute for Cancer, Stem Cell Biology and Medicine" [online], http://mednews.stanford.edu/stemcellQA.html.
4. "Islet Transplantation in Seven Patients with Type I Diabetes Mellitus Using a Glucocorticoid-Free Immunosuppressive Regimen," *New England Journal of Medicine* 343 (July 27, 2000): 230–38.
5. "Pancreas Stem Cells for Diabetes," BBC News, August 23, 2004 [online], http://bbc.co.uk/1/hi/health/3584090.stm.

Part 3
Moral Issues

Editors' Introduction

Some people simply think that stem cell research is wrong, absolutely and completely. The conservative commentator Michael Novak is one. He takes a Kantian position, arguing that one should never destroy a human being for the benefit of another, although note that he and those whom he mentions favorably think that the notion of human being goes right back to the very moment of conception or its equivalent. Others would obviously disagree, including the ethicists drawn on by the biotech company Geron, who argue that in properly controlled circumstances one can indeed move forward ethically on stem cell research.

Note that like many consequentialists, these people are concerned about overall good, basing their recommendations at least in part on global concerns. For them, part of the ethical justification of such research comes from the benefits that would (or should) be received by the many and not just the few. Also note that to some extent they try to side step the issue of the status of the fetus by arguing that if the fetus is already available (that is through abortion for unrelated reasons), then this opens the way for its use. In other words, they try to brick off the moral issue of killing the fetus from that of using the fetus. Obviously, this is a crucial move and one doubts that it would be acceptable to everyone. These are some of the issues raised by John Robertson in his detailed discussion, one that concludes that one can use embryos for stem cell research—including embryos discarded in fertility treatments and even those created specially for research and therapeutic purposes. Noted specialists in bioethics Glenn McGee and Arthur Caplan take up the debate and they argue vigorously that too much discussion of these issues has been based on politics first and ethics second, whereas truly the order should be reversed. They argue that, on balance, the worth of the embryo does not equal that of the human suffering from disease and hence it is morally justifiable to sacrifice the former for the sake of the latter. Whether they make their case successfully is for you to judge.

We also have two excellent articles by Holm and Marquis both of which argue that there is something wrong with stem cell research. Holm tackles the positive argument for stem cell research and claim to show that they fail to get the conclusions they need. Marquis, who is famous for his secular argument against abortion, works toward providing a nonreligious argument for banning stem cell research.

Global and social justice are also important moral concerns in the stem cell debate. For this we have two articles, one that is concerned with the issue of who will benefit from stem cell research and stem cell banks. Will it be privileged white America who will be the primary beneficiary of stem cell research? If so, does this provide good reason to think that money should be put into other kinds of programs that will promote social justice? And of course there is always the tension between individual autonomy and social justice.

QUESTIONS TO THINK ABOUT

(1) How are the stem cell debate and the abortion debate similar and dissimilar? Could a person have one view about abortion and a separate view about stem cell research and still be morally consistent?
(2) What role should government play in creating a just society, and

does that affect the kind of decisions that should be made with respect to stem cell research?

(3) Do potential people have actual, current rights? If so, what kind of duty do we have to grant those rights? If they don't have rights, then do they still have moral claims against actual people, and what would those be?

(4) If stem cell research does actually produce cures that will curb the suffering of millions of people, does it justify the use of stem cells that could (in a very broad sense) become a person?

(5) Because science is an international activity and it cannot be stopped by any particular country, what kind of responsibility does science have to police itself? What kind of behavior should scientists engage to act morally? Could a virtue ethics like Aristotle's help in this kind of area?

11
The Stem Cell Slide: Be Alert to the Beginnings of Evil

Michael Novak

I wish I could say that the president's speech to the nation on stem cells was as good as I had hoped. It was in many ways a wonderful speech: deeper and more philosophical than I have ever heard a president deliver, unusually balanced and fair in presenting opposing arguments, and clear in delineating both his own decision and the reason for it. It was, in addition, heartfelt and compassionate toward all families who have members suffering from awful diseases or disabilities. I can even see how the president convinced himself, at the end, that he had found a ray of daylight through the opposing arguments, and arrived at a moral decision that seemed to him sound, and also politically defensible. During the last few months, I have heard many persons who think they are very smart lay out their arguments on this question. Not one of them did as thorough, many-sided, fair, and clear a job as President Bush did in his speech.

At the end, though, my heart sank. The president tried to maintain a position of principle, but what he ended up doing, despite his best effort, was giving away the principle. He put the Full Faith and Credit of the U.S. government behind the principle of using human beings as a means, albeit for noble ends. He offered a reason for doing this: The stem cells for whose use in experimentation he commits federal resources come from embryos already destroyed. Why not bring good out of evil, he argues, by now using these stem cells, which will otherwise be wasted, to search for cures for awful diseases? The outcome is not certain, but at least it's noble to try. This is a lovely and tempting thought. The problem is that when the source of stem cells runs out—soon—then those on the other side will demand more stem cells from more embryos. The demand for usable stem cells will swell enormously. This is particularly true if good experimental results are obtained. But it will even be true if they aren't: Look, partisans will say, you were too stingy, too narrow. You have ceded the principle, so now give up more of the specifics. The glittering utopia of science beckons just ahead.

Be alert to the beginnings of evil. It never comes under the appearance of evil, but always under the appearance of the beautiful, the promising, the idealistic, the pleasant. "Stop it in its beginnings," the ancient principle runs: the sooner, the easier.

Politically, the decision may play very well among a substantial majority. It is already clear that those on the left who you hope will attack it are attacking it, which will only reinforce those among right-to-lifers who accept the president's obvious good will, often deeply moving words, clearly articulated argument, and patent depth of feeling. But I deeply fear the immense battles that lie ahead, and the gathering of heartened foes, who will very quickly sniff out the weak point and pry its own inner logic with all their force. It will take almost superhuman strength now for the president to hold the new position he has moved into, having surrendered the strongest ground.

That ground was a philosophical one, not a theological one, a ground born of reason rather than of faith. One of its classic articulators was Immanuel Kant. The president himself alluded to it in his speech, in the line about not using human beings as means for even the noblest ends. You must never use a human being as a means for even the noblest of ends. To use stem cells obtained by killing living human beings in their embryonic stage is still using them as a means. It is not enough to say that the wicked deed has already been done—that the embryos have already been killed. The purpose of that killing was to obtain the stem cells. One ought not to implicate oneself in that process, not even for the noblest and most beautiful ends.

One especially ought not to implicate the United States, which Hannah Arendt once called humankind's noblest experiment. For this nation began its embryonic existence by declaring that it held to a fundamental truth about right to life endowed in us by our Creator. The whole world depends on our upholding that principle.

We human beings very easily reason ourselves into taking positions that end up having the most tragic of consequences, positions of which we would never have approved had we seen those consequences at the time. For the fruit of the tree of knowledge over yonder appears to be very sweet, and we feel sure that if we eat of it, then happy endings (fit for a god) will result. Those endings have always turned to sulphur in our cheeks.

The fatal mistake often comes as a result of unexamined moral sentiments: affects and feelings that serve as moral guideposts without submitting to interrogations by reason. "I feel comfortable with that," President Bush—like President Clinton, indeed like just about everybody else in this fair land—is wont to say. It is as if Americans were ashamed to say that they reached a considered intellectual judgment, independently of their feelings. "I feel comfortable with that" seems itself to be more comfortable than "That's what I've reasoned to."

And this should cause us great uneasiness, because very often in the moral life, our feelings and sentiments are horrible guides to right action. It sometimes feels like sheer Hell to have to do the right thing, and most terribly *un*comfortable. For instance: When individuals in Nazi Europe made the personal decision to join the Resistance, they often did so feeling the most awful dread, sick in their stomachs about the prospect of being hunted out like animals, tortured, and killed in barbaric ways (on meat hooks, say). They could not afford to listen to their feelings. (Considerations of just this sort led Karol Wojtyla to abandon the philosophy of the moral sentiments he learned from Max Scheler, and to search for a philosophy that drove deeper to intellectual principle and a strong moral will.)

One thing this debate is showing the nation is the difference between those who can grasp, and be swayed by, intellectual principles and those who need warmer, fuzzier comforts for their senses, imaginations, and sentiments. Thus, when Senator [Orrin] Hatch finds it easier to imagine life in the womb than in a clump of cells in a petri dish, he is not wrong about what is occurring in his imagination and sentiments. But those means of perception do not dig deep enough into the *what-IS-it*? of those cells. To do that, cold intellect must go to work, beyond the comfort zone of imagination and sentiments.

And those cells are a living human being—at a very early stage, to be sure, but unmistakably human: not a rat, or a cocker spaniel, or a fish. That's precisely what makes them so desirable to researchers.

Desirable: There's a key word. It is transparent, now, how hungry researchers are to get their hands on these embryos. And perhaps on the money and the fame that beckon just ahead. As Mr. Dooley reminds us, when a fella says, 'tain't the money, it's the principle—it's the money. Suddenly, even great scientists are forgetting the basic biology still being taught in the latest textbooks: that the first appearance of *human* beings is in the fertilized egg embryo. Suddenly, scientists are fudging: "Well, really, the human being comes later." Let us show no respect for the human being in embryo, not now. Postpone respect until later—not for this class of human beings.

The role of *desire* is palpable. Researchers *want* this research. The political class *wants* this research. People are making themselves believe, without evidence, and despite many warnings signs to the contrary, that there *will* be glitteringly good results from these experiments, and *only* good results. Desire is getting miles ahead of cool judgment. The extremely plausible horrors of the future are being systematically kept out of the imagination.

I have enormous sympathy for Christopher Reeve, and profoundly hope that, by some miracle of medicine or grace, he is suddenly healed. Yet I am also disappointed by how earnestly one whom I still think of as Superman wants to have other human beings killed, so that he might be cured. He has also plaintively mentioned that some who now don't want embryos destroyed never said a word on their behalf earlier. But in fact, some fifteen years ago, the pope and Cardinal Ratzinger took a lot of heat for their principled opposition to in vitro fertilization—that is, to the very techniques of "clumps of cells in a petri dish" and "cells in refrigerators," followed by the wanton destruction of faulty or unwanted embryos, that so many now deplore as less than human. Disagree with their moral conclusions, then or now, if your own mind leads you to. But do not say that they did not grasp, earlier than most, the intellectual principles that are now unfolding before our sentient eyes, and our recoiling imaginations, and our resistant sentiments. Do not say that everyone was silent.

We are testing a great political principle: whether a nation of the people, by the people, and for the people can form great public decisions through open public argument, reflection, and considered choice—or must forever form them by passion and bias and desire and emotional herding.

The president, though he stumbled in his moral reasoning, conceived and executed a shrewd enough political stroke to have temporarily disarmed his foes, won some time, and earned sufficient public standing to lead the nation through a great new era in our history. There was once, in the eighteenth century, a "new science of politics" and , later, a "new science of economics," to both of which our Founders contributed their share of innovations. We are now engaged in learning a "new science of morals," or more exactly a new science of public moral management.

This is the worthiest of tasks for a free society, because what is public freedom for, if not for well-argued and wise moral action? What is the point of political liberty and economic liberty, if we are to live like less-than-human animals?

12 Research with Human Embryonic Stem Cells

Ethical Considerations

Geron Ethics Advisory Board (Karen Lebacqz [Chair], Michael M. Mendiola, Ted Peters, Ernlé W. D. Young, and Laurie Zoloth-Dorfman)

RESEARCH WITH EMBRYONIC STEM CELLS

An Ethics Advisory Board, whose members represent a variety of philosophical and theological traditions with a breadth of experience in health care ethics, was created by Geron Corporation in 1998. The board functions as an independent entity, consulting and giving advice to the corporation on ethical aspects of the work Geron sponsors. Members of the board have no financial interest in Geron Corporation.

The Geron Ethics Advisory Board is unanimous in its judgment that research on hES cells can be conducted ethically. In order for such research to be conducted ethically in the current context, some conditions must pertain. In addition, further public discourse will be needed on a range of ethically complex questions generated by this research. The conditions are:

1. The blastocyst must be treated with the respect appropriate to early human embryonic tissue.
2. Women/couples donating blastocysts produced in the process of in vitro fertilization must give full and informed consent for the use of the blastocysts in research and in the development of cell lines from that tissue.
3. The research will not involve any cloning for purposes of human reproduction, any transfer to a uterus, or any creation of chimeras.
4. Acquisition and development of the feeder layer necessary for the growth of hES cell lines in vitro must not violate accepted norms for human or animal research.
5. All such research must be done in a context of concern for global justice.
6. All such research should be approved by an independent Ethics Advisory Board in addition to an Institutional Review Board.

On 5 November 1998 Geron Corporation announced that scientists working in collaboration with Geron had succeeded in establishing cell culture lines of human embryonic stem (hES) cells. Because these cells are considered pluripotent (capable of being the precursor to a variety of human cell types) and immortal (sustainable in culture and reproducing themselves indefinitely), they represent a major breakthrough in scientific research, with potential for significant advances in tissue transplantation, pharmaceutical testing, and embryology.

Prior to the November announcement, the Geron Ethics Advisory Board developed "A Statement on Human Embryonic Stem Cells"[1] as a set of guidelines for hES research. [See above.] This essay provides an expansion and elaboration of the particular warrants and moral reasoning for that statement.

The EAB did not offer a *carte blanche* approval of research on hES cells undertaken by Geron or any other entity. The board unanimously affirmed that such research can be undertaken ethically, contingent upon meeting a range of qualifying conditions. The initial work of the board has been to specify such conditions; its continued work will consist of assessing Geron's developing research in light of them. Hence here we also include some preliminary reflections on ethical issues in human embryonic germ cell (hEG) research.

Moreover, the EAB perceived the need for urged continued public discussion of the complex ethical issues emerging form such research. Thus the statement and this companion essay should be seen not only as an initial clarification of the EAB's own position, but also as an effort to contribute to and invite that public discourse. We enumerate some specific questions for further reflection at the end of this essay.

1. "The Blastocyst Must Be Treated with the Respect Appropriate to Early Human Embryonic Tissue."

The creation of hES cells involves isolation of cells from the blastocyst.[2] The blastocyst consists of an outer cellular layer, which would develop as the placenta, and an inner cell mass, which would develop as the body of the fetus. The outer layer is dissolved and the resulting mass of cells is used for research. Thus a central ethical issue is the moral status of the blastocyst.

To raise the matter of "moral status" is to ask, Does a given entity possess the requisite qualities or characteristics that entitle it to moral consideration and concern? "Moral status" thus functions as a threshold idea: entities with moral status should be treated in a manner differently form Entities without that status. The EAB affirms that the blastocyst has moral status and hence should be treated with respect.[3]

What sort of moral status does the blastocyst have? This question has riveted political, religious, and ethical attention, and profound and substantial disagreement is based not only on contending biological interpretations but also on deeply held philosophical and theological considerations. Some have argued for conception as the relevant consideration, others for the development of the "primitive streak" (the precursor to the spinal cord of an individual fetus) as a defining moment, and some for utilizing implantation as the crucial threshold for moral status.[4]

Reviewing the complex literature on this topic, Ted Peters, following Daniel Callahan, distinguishes three basic schools of thought.[5] The *genetic school* locates the beginning of human personhood, and thus claims of moral status and dignity, at the genetic beginning—that is, at conception, at the point where one's individual genome is set. Here, a criterion for moral status (human genetic heritage) is linked to a particular point in human life (conception). The *developmental school*, while granting that human life begins at conception, bolds equally that human *personhood*—and hence *full* moral status—is a later development. Here, moral status is understood developmentally: as the conceptus develops from blastocyst to fetus and beyond so too does moral status grow (although proponents differ on when exactly the threshold of moral status is reached). The *social consequences school* shares with the developmental school the belief that human personhood is a process and an achievement over time. Advocates deny, however, that personhood is achieved at any particular moment. Rather, "personhood" is a matter of definition rather than biological fact, based on socially constructed norms.

In its work, the National Institutes of Health Human Embryo Research Panel focused less on the time when moral status might be acquired and more on the *criteria* for its determination. The panel noted two broad approaches in the debates: one proposes a single criterion as constitutive of moral personhood, while a second, "pluralistic," approach emphasizes a number of different, interacting criteria. As the panel noted

> Among the qualities considered under a pluralistic approach are those mentioned in single-criterion views: genetic uniqueness, potentiality for development, sentience, brain activity, and degree of cognitive development. Other qualities often mentioned are human form, capacity for survival outside the mother's womb, and degree of relational presence (whether to the mother herself or to others). Although none of these qualities is by itself sufficient to establish personhood, their developing presence in an entity increases its moral status until, at some point, full and equal protectability is required.[6]

The panel proposed the pluralistic approach as the more adequate of the two, with moral status (and hence protectability) understood developmentally, culminating at birth in full and equal personhood.[7]

Drawing upon this wealth of philosophical and theological reflections and situating ourselves relative to it, the EAB affirmed our understanding of moral status as developmental and consonant with the pluralistic approach. This developmental view is in accord with Jewish tradition,[8] with the views of many Roman Catholics[9] (although not the Vatican), and with the majority of Protestant traditions as well as with legal traditions that provide different protections at different stages of fetal development.

We hold that a fundamental principle of respect for human life applies at all states of human development. The developmental view that we affirm does not mean that the principle of respect can be ignored; it means that the principle requires different considerations and entails different obligations at different developmental states. For example, Lisa Sowle Cahill argues that "Few doubt that there exists from conception some form of 'human life' in the literal sense; the crucial question is whether from conception or at any subsequent time during pregnancy that life deserves the same respect and protection due an infant."[10] Once there is evidence of capacity for sensation, for example, respect requires minimization of pain. In this very early embryonic tissue there is no capacity for sensation; thus minimization of pain does not apply. Rather, early embryonic tissue is respected by ensuring that it is used with care only in research that incorporates substantive values such as reduction of human suffering.[11] We believe that the purposes of the hES research—its potential to contribute to fundamental knowledge of human development and to medical and pharmaceutical interventions to alleviate suffering—provide such substantive values.

A second source of cells is human embryonic germ (hEG) cells derived from gamete ridge tissue removed from early fetal tissue following elective abortion. These cells have been cultured using similar but not identical methods as are used for the hES cells, and may have both properties of pluripotency and immortality. However, this research raises ethical questions distinct from those in the use of the early blastocyst. The tissue is taken (much as cadaver organs might be taken) from an aborted fetus within the first eight weeks after conception. At stake in this debate is whether licit use can be made of tissue collected after abortion in a society in which the act of abortion is seen by some as murderous and by others (and by law) as acceptable.

The EAB cannot resolve the contentious abortion debate. We are developing guidelines for hEG research. Preliminary reflections suggest at least the following concerns: First, all agree that the demise of the fetus is not caused by the research procedures. Second, the moral obligation to save life may be a sufficiently strong warrant to justify certain uses of the tissue of the dead and hence to support such research. Third, the tissue of the dead must be used with respect. Respect for tissue taken from a dead fetus would take into account the need for closure, grief, and ritual that families might have in these cases. Respect would include the confidential and dignified handling of the tissue when collected and used. Finally, issues of informed consent would apply with the same stringency required in the case of hES research (see below).

2. "Women/Couples Donating Blastocysts Produced in the Process of *In Vitro* Fertilization Must Give Full and Informed Consent for the Use of the Blastocysts in Research and in the Development of Cell Lines from That Tissue."

Human embryonic stem cells are derived from embryos produced from clinical purposes and then donated for research purposes. Hence of crucial ethical significance is the character and quality of the consent given by women and couples to such donation. As with the issue of moral status, the ethical and legal literature on consent and refusal is voluminous.[12] What is needed for valid consent to donate embryonic tissue for research purposes?

Tom Beauchamp and James Childress argue that the core meaning of informed consent is an *autonomous authorization* of a medical intervention or involvement in research. An informed consent occurs "if and only if a patient or subject, with substantial understanding and in substantial absence of control by others, intentionally authorizes a professional to do something."[13] Thus informed consent or refusal is not simply a matter of what the professional should disclose to the patient, but is a matter of ensuring that the conditions necessary for autonomous authorization are met—i.e., that consent is intentional, with substantial understanding, and without controlling influences. Authority for participation in research, expressed in an informed consent, resides with and is to be exercised by the subject.

Consent to utilize embryos for *clinical* purposes if IVF does not therefore suffice as consent for their use in research and cell-line development. Explicit consent must be elicited for such use. We concur with Arthur Caplan that "when research is the goal, whether for profit or not, those whose materials are to be used have a right to know and consent to such use."[14] Other commentators agree that fairness and respect for persons dictate that explicit information be provided to patients regarding the use of their tissues and cells, especially when those may be used for commercial purposes.[15] For these reasons, the EAB determined that women/couples donating embryos for this research should understand clearly the nature of the research and also should understand whether there are commercial implications and if so, whether they hold any proprietary rights in the tissue lines developed from embryonic cells.[16]

Moreover, we believe that the context of embryo donation within the process of IVF renders the need for careful consent even more important. The IVF process is often physically painful, emotionally burdensome, and financially costly. These factors may make IVF patients particularly vulnerable.[17] Such possible vulnerability demands careful and consistent efforts on the part of researchers not to exercise a controlling or coercive influence over these patients. A recent study of patients who consented to participate in therapeutic research found that "physicians' recommendations were . . . powerful factors influencing patients' decisions to become research subjects."[18] Authors of the study recommended that research ethics "be enriched with a sensitivity to the profound trust participants place in researchers and the research enterprise."[19] The context of a study regarding participation in therapeutic research is clearly different from the context of IVF embryo donations for hES cell research. In the latter context, the consent is "twice removed": women/couples are not consenting to therapeutic research, nor does the research involve their own bodies or persons. Nevertheless, even in the IVF context some measure of trust in the researchers may be involved and this possibility raises a warning regarding the ways in which trust can be manipulated to researchers' advantage, thus possibly leading to undue influence over subjects. Moreover, while this research does not involve their own bodies, it does involve entities that they may well consider as potential progeny, a factor that must be considered in this research.

Informed consent also requires "substantial understanding." Hence the EAB requires certain informational elements in the context of embryo donation for hES cell research. First, as the derived hES cells are of significant market potential, donors should understand those potential market implications. Second, they should be advised as to whether there are any proprietary rights in the tissue.

The achievement of substantial understanding requires a process that builds trust over time, an admittedly difficult requirement yet one that will be critical in all stages of this research for it to be carried out ethically.[20] Adequate time for eliciting and answering donor questions is necessary, as is ensuring that donors are genuinely free to refuse as well as to consent.

In the course of our deliberations, we reviewed an exemplary form used in soliciting consent for embryo donation for research.[21] This form states explicitly: "cells that may be derived from embryos donated for this research could have clinical and commercial value in the event that the study is successful." It further specifies:

> This research will not benefit your clinical care, and you will not benefit financially from it. The physicians/clinicians involved in your care will not benefit financially from this study. The investigators conducting this study could benefit financially from clinical or commercial values that may result from it. The cells derived in this study may be shared with Geron Corporation, located in Menlo Park, CA, as part of the study. Geron Corporation may benefit financially from the development and clinical use of the cells derived in this study.

Such explicit statements embody well the kinds of financial disclosure that a consent form should contain.

3. "The hES Research Will Not Involve Any Cloning for Purposes of Human Reproduction, Any Transfer to a Uterus, or Any Creation of Chimeras or Human-Animal Hybrids."

Corporate representatives have stated clearly to the EAB that human reproduction was not a goal, purpose, or intent of Geron's research on hES (and hEG) cells. Any effort to produce a living being out of this research would raise a host of ethical issues, in particular the risk of harms to potential offspring,[22] to parenting women and couples, and to the human community itself through genetic manipulation and transmission. Because Geron is not engaged in these activities, and has made clear its intention not to do so, the EAB did not take them up for extended assessment in its deliberations. Should Geron consider initiating any such activities, the EAB will undertake the necessary ethical analysis.

4. "Acquisition and Development of the Feeder Layer Necessary for the Growth of hES Cell Lines *In Vitro* Must Not Violate Accepted Norms for Human or Animal Research."

To keep hES cells in an undifferentiated state, they are cultured and maintained on layers of nutrients called "feeder layers." Currently, irradiated mouse embryonic fibroblast feeder layers are utilized, but it is possible that other tissues would be use in the future. The acquisition and development of the feeder layers must be in accord with norms for research appropriate to the source from which the feeder layers are drawn.[23]

Geron's use of mouse embryonic cells would seem to fall well within a judgment of ethical use of animal tissue. The research holds great potential therapeutic value. In addition, mouse embryos fall below the threshold of sentience (the ability to experience pain) or of other capacities of organic animal being and activity. Nevertheless, we mandate continued attention to animal welfare.

5. "All Such Research Must Be Done in a Context of Concern for Global Justice."

One of the primary justifications of hES research is beneficence based: its therapeutic potential to alleviate human suffering and to promote the health and well-being of human populations. However, to justify a practice on the basis of its benefits makes moral sense only if people in need actually have access to those benefits. Hence the justification gains credibility only when it is wedded to a commitment to justice, rooted in "a recognition of our sociality and mutual vulnerability to disease and death."[24] *The EAB considers concerns about social justice in public health to be of overriding importance.* Thus in the EAB's judgment, it is morally paramount that research development include attention to the global distribution of and access to these therapeutic interventions.

Two features of Geron's research render this commitment to just access particularly challenging. First, the research is undertaken in the private sector—in the context of market forces, patenting of products, interests of shareholders and investors, and a consideration of profit. These varied interests may compete with—but should not override—a concern for equitable access.[25] Second, the research is highly technological and expensive, as well as under the proprietary rights of a U.S. company. How to ensure adequate access for insured, underinsured, and noninsured patients in the United States, let alone on a global basis, will be an ethically and financially challenging task. The EAB will continue to work with Geron on these matters.

6. "All Such Research Should Be Approved by an Independent Ethics Advisory Board in Addition to an Institutional Review Board."

At its own initiative, Geron established the EAB. Given the kinds of ethical issues already emerging from hES research and those yet to come, the EAB reaffirmed the necessity of an independent board for ethical analysis and consultation. In addition, the research undertaken should receive IRB review and approval in order to protect subjects involved.

ISSUES REQUIRING FURTHER DELIBERATION

As indicated above, the EAB wishes to generate public discourse on a wide range of issues related to emerging research arenas. Among those are some that surfaced during our discussion but are not reviewed in this essay.

1. Who should exercise control over the disposition of fetal or embryo tissue? In our discussion of consent, we refer to both "women" and "couples." The appropriate locus of consent/refusal may be disputed.

2. What is the proper relationship of ethicists to proprietary companies? Who should constitute an ethics board, and who should serve on it? Under what conditions (e.g., remuneration, stock options, etc.)?[26]

3. As noted above, how can difficult issues of global justice and fair distribution be handled in research involving private enterprise?

4. What is the role of consensus in a society that is both pluralistic and often deeply divided over appropriate norms? How can we develop appropriate language for public debate and decision making while remaining respectful of differences and accountable to substantive moral disagreements?[27]

NOTES

1. The statement was drafted in September 1998 by Karen Lebacqz, revised by EAB, and was finalized on October 20, 1998.

2. The blytocyst is approximately 140 cells, fourteen days postconception.

3. Michael M. Mendiola, "Some Background Thoughts on the Concept of 'Moral Status' Relative to the Early Embryo" (Background paper for the EAB).

4. Jacques Cohen and Robert Lee Hotz, "Toward Policies Regarding Assisted Reproductive Technologies," in *Setting Allocation Priorities: Genetic and Reproductive Technologies*, ed. Robert H. Blank and Andrea Bonnicksen (New York: Columbia University Press, 1992), pp. 228–29.

5. Ted Peters, *For the Love of Children* (Louisville, Ky.: Westminster/John Knox Press, 1996), pp. 96–100.

6. National Institutes of Health, *Report of the Human Embryo Research Panel* (Bethesda, Md.: National Institutes of Health, 1994), p. 49. Mary Anne Warren offers a critique similar in structure, arguing for a "multi-criterial" view of moral status, instead of "unicriterial" approaches. Mary Anne Warren, *Moral Status: Obligations to Persons and Other Living Things* (Oxford: Clarendon Press, 1997).

7. National Institues of Health, *Report of the Human Embryo Research Pannel*, pp. 49–51. Cohen and Holtz echo this position: "New life does not appear suddenly; it is created gradually, with each new phase differing from that which preceded it." Cohen and Holtz, "Toward Policies Regarding Assisted Reproductive Technologies," p. 230.

8. Laurie Zoloth-Dorfman, "The Ethics of the Eighth Day: Jewish Bioethics and Genetic Medicine" (Background paper for EAB).

9. See, for example, Thomas A. Shannon and Allan B. Wolter, "Reflections on the Moral Status of the Pre-Embryo," *Theological Studies* 51 (1990): 603–26.

10. Lisa Sowle Cahill, "Abortion," in *The Westminster Dictionary of Christian Ethics*, ed. James F. Childress and John Macquarrie (Philadelphia: Westminster Press, 1986), p. 3.

11. Ernlé W. D. Young, "The Moral Status of Human Embryonic Tissue" (Background paper for the EAB). The EAB holds that "substantive" values are those supported by prima facie moral duties such as nonmaleficence.

12. See Ruth R. Faden and Tom L. Beauchamp, *A History and Theory of Informed Consent* (New York: Oxford University Press, 1986), for a good overview of the historical development and philosophical treatment of informed consent in the clinical and research contexts.

13. Tom L. Beauchamp and James F. Childress, *Principles of Biomedical Ethics*, 4th ed. (New York: Oxford University Press, 1994), p. 143.

14. Arthur L. Caplan, "Blood, Sweat, Tears, and Profit: The Ethics of the Sale and Use of Patient Derived Materials in Biomedicine," *Clinical Research* 33, no. 4 (1985): 451.

15. See, for example, George J. Annas, "Outrageous Fortune: Selling Other People's Cells," *Hastings Center Report* 20, no. 6 (1990): 36–39; also George J. Annas, "Whose Waste Is It Anyway? The Case of John Moore," *Hastings Center Report* 18, no. 5 (1988): 37–39.

16. The most celebrated legal case dealing with the issue of use of tissue without consent is that of *Moore v. Regents of the University of California* (51 Cal 3rd 120, 1990). For an extended discussion, see E. Richard Gold, *Body Parts: Property Rights and the Ownership of Human Biological Materials* (Washington, D.C.: Georgetown University Press, 1996).

17. Judith Lorber has argued that IVF may also involve a coercive element for women when undertaken to treat male infertility. Such women may have to strike a "patriarchal bargain" to maintain a relationship and have a child within the constraints of monogamy, the nuclear family, and the valorization of biological parenthood. Judith Lorber, "Choice, Gift, or Patriarchal Bargain? Women's Consent to In Vitro Fertilization in Male Fertility," in *Feminist Perspectives in Medical Ethics*, ed. Helen Bequaert Holmes and Laura M. Purdy (Bloomington: Indiana University Press, 1992), pp. 169–80.

18. Nancy E. Kass et al., "Trust: The Fragile Foundation of Contemporary Biomedical Research," *Hastings Center Report* 26, no. 56 (1996): 26.

19. Ibid., p. 27.

20. Mark G. Kuczewiski, "Reconceiving the Family: The Process of Consent in Medical Decisionmaking," *Hastings Center Report* 26, no. 2 (1996): 32.

21. The form was developed by Roger Pedersen, Ph.D., of the Department of Obstetrics, Gynecology, and Reproductive Sciences, University of California, San Francisco, and is used with permission.

22. It is for this reason that the NIH Human Embryo Research Panel distinguished between research on embryos intended for and not intended for transfer. Research on the former must include consideration of potential harms to the future child—a concern not raised by the latter. (National Institutes of Health, *Report of the Human Embryo Research Panel*, pp. 51–52.) It should also be noted that hES cells are derived cells and are not the cellular equivalent of an intact embryo. Even if transferred, hES cells would not form an embryo because they lack other cells necessary for implantation and embryogenesis.

23. The current usage of mouse embryo cells raises ethical concerns about the use of animals in hES research. A number of criteria have been proposed to weigh ethical versus unethical uses of animals in biomedical research. However, as Strachan Donnelley notes, no single, unambiguous

standard or guideline exists for assessing each and every use of animals in science. Ethical attention must be directed to the specific purposes, types, and contexts of animal use. See Strachan Donnelley and Kathleen Nolan, eds. "Animals, Science, and Ethics" [Special Supplement], *Hastings Center Report* 20, no. 3 (1990): S11–S12.

24. Larry R. Churchill, *Rationing Health Care in America: Perceptions and Principles of Justice* (Notre Dame, Ind.: University of Notre Dame Press, 1987), p. 135.

25. Zoloth-Dorfman, "The Ethics of the Eighth Day."

26. Members of the EAB receive a very modest honorarium for the time spent in scheduled meetings with Geron staff and officers. We are not compensated for time spent in research, writing, or other conversations with Geron, nor does any of us hold stock in the corporation.

27. See, for example, Alta Charo, "The Hunting of the Snark: The Moral Status of Embryos, Right to Lifers, and Third World Women," *Stanford Law and Policy Review* 6, no. 2 (1995): 11–38.

13
Ethics and Policy in Embryonic Stem Cell Research

John A. Robertson

Mammalian tissue and organs derive from pluripotent embryonic stem (ES) cells present at the blastocyst stage of embryo development. Embryonic stem cells were first isolated from the inner cell masses of mouse blastocysts in the early 1980s. In November 1998, a team of researchers at the University of Wisconsin and a team at Johns Hopkins University published reports of the first successful isolation and culturing of human ES cells.[1] Because ES cells are capable of self-renewal and differentiation into a wide variety of cell types, the ability to grow them in renewable tissue cultures could have broad applications in research and transplantation.

The research uses of human ES cells include in vitro studies of normal human embryogenesis, abnormal development, human gene discovery, and drug and teratogen testing. Potential clinical applications are as a renewable source of cells for tissue transplantation, cell replacement, and gene therapy. For example, if human ES cells could

John A. Robertson, "Ethics and Policy in Embryonic Stem Cell Research," *Kennedy Institute of Ethics Journal* 9, no. 2 (1999): 109–36.

be directed to differentiate into particular tissues and immunologically altered to prevent rejection after engraftment, they could treat or cure thousands of patients who now suffer from diabetes, neurodegenerative disorders, heart disease, and other illnesses.

The growth of human ES cells in culture is a first but necessary step toward development of cell replacement or regeneration therapies. Future work will have to determine how to obtain human ES cells efficiently and reliably from the inner cell mass of human embryos or from primordial germ cells, grow them in culture, and then identify the growth factors that will direct them to differentiate into cells of particular types to produce the large enough number of pure cells that will be necessary for transplantation.[2] Finally, clinical research using ES-derived cells will be needed to determine under what conditions they are therapeutic for the many conditions that they potentially could treat. Of major importance will be tailoring stem cells genetically to avoid attack by a patient's immune system.

The pace of stem cell research is likely to be affected by ethical and legal concerns beyond the issues of animal data, safety, informed consent, and IRB review that arise with any kind of clinical or human subjects research. The ability to regenerate vital tissues through ES cell technology could eventually transform our understanding of aging and personal identity, as well as place heavy burdens on medical care and social security systems from new ways to extend adult life. Some persons may have serious moral qualms about ES cell research for these reasons alone, and question whether public research support for cell regeneration technologies should be provided at all.

More immediately, ethical and legal issues arise with ES cell research because ES cells have to be derived from aborted fetuses or from the inner cell mass of human preimplantation embryos. Although the ethical and legal issues involved in the retrieval of cells from aborted fetuses and embryos differ in some respects, both sources of ES cells raise concerns about respect for human life at its earliest stages and the extent to which such life may be used or destroyed to provide cells or tissue for research or therapy.

Such questions are highly controversial. Unless ES cells could be obtained from spontaneously aborted fetuses or from preimplantation embryos without destroying them, which appears to be highly unlikely, questions of ES cell research and therapy are likely to rekindle bitter controversies over fetal tissue transplantation and embryo research. Research with more differentiated stem cells, such as mesenchymal stem cells, which already can be directed in vitro to become cartilage, bone, fat, tendon, or stroma, will raise fewer ethical and legal problems because those cells can be obtained from consenting persons.[3] However, those cells will not meet the need for many kinds of tissue that can be derived only from pluripotent stem cells. A conflict between respect for the first stages of human life and the persons who will benefit from such research appears to be inevitable.

Legally, research or therapy with fetal tissue derived from induced abortions is now permissible with federal funds. Current federal law, however, prohibits the use of federal funds for the "creation of a human embryo" for research purposes or for "research in which a human embryo or embryos are destroyed, discarded or knowingly subjected to the risk of injury or death."[4] This law broadly defines the term "embryo" to include those that are "derived by fertilization, parthenogenesis, cloning, or any other means from one or more human gametes or diploid cells."

The federal prohibition on the funding of embryo research does not bar ES cell research with private funds in most states, but it slows the progress of such research and creates practical barriers for scientists working in institutions that receive federal research funds.[5] It also leaves researchers without a clear set of ethical or regulatory guidelines for ES cell research with human embryos. With ES cell research now ripe for further advances, and the National Bioethics Advisory Commission (NBAC) and other policymaking bodies considering the issue, it is time to assess the ethical limits on the use of aborted fetuses or IVF-created embryos to obtain ES cells for research and therapy.

ARE HUMAN EMBRYONIC STEM CELLS EMBRYOS?

Ethical questions arise about ES cell research because ES cells must first be derived from human aborted fetuses or preimplantation embryos. It should be clear, however, that ES cells are not themselves embryos because, although they are *pluripotent* in that they could develop into any cell or tissue of the body, they are not *totipotent*. They are not capable of forming a new individual, as a fertilized egg or single cell taken from a four-cell embryo might if cultured in vitro and placed in a uterus. Culture of ES cells followed by placement in the uterus would not result in the implantation of an embryo and eventual birth of a child.

General Counsel Harriet Raab of the Department of Health and Human Services relied on this distinction in a legal opinion to the director of the National Institutes of Health on the acceptability under current federal law of research with ES cells. She concluded that the "statutory prohibition on the use of funds appropriated to HHS for human embryo research would not apply to research utilizing human pluripotent stem cells because such cells are not a human embryo within the statutory definition."[6] That definition refers to embryos as "any organism, not protected as a human subject under 45 CFR 46."[7] Referring to a standard dictionary meaning of organism as "an individual constituted to carry out all life functions," Counsel Raab stated:

> Pluripotent stem cells are not organisms and do not have the capacity to develop into an organism that could perform all the life functions of a human being—in this sense they are not even precursors to human organisms. They are, rather, human cells that have the potential to evolve into different types of cells such as blood cells or insulin producing cells.[8]

Counsel Raab's conclusion that ES cell research per se is not covered by the current federal ban on embryo research appears to be legally correct, for ES cells, as she notes, are not themselves embryos nor are they capable of becoming embryos.[9] Her opinion, however, rests on a technical distinction between destructive removal of the ES cells from embryos, which is not permitted with federal funds, and the research that then occurs with those cells after removal. Under Raab's ruling, federal research funding could be provided for research with ES cells as long as the actions involved in removing the ES cells were not federally funded. This technical distinction did not sit well with pro-life members of Congress, who protested that it "would violate both the letter and the spirit of the federal law."[10] Several members of the National Bioethics Advisory Commission also found Raab's conclusion to be "disingenuous" and "legalistic," and objected to authorizing federal support for ES cell research on the basis of that distinction alone.[11] Others might find the distinction useful because it would permit some federal funding of ES cell research without requiring a change in current federal law.[12]

TWO KINDS OF COMPLICITY

Criticism of the HHS Counsel's position rests on the belief that the derivation and use of ES cells cannot be morally or legally separated. If the initial derivation of ES cells from embryos or aborted fetuses is immoral, as many persons believe, then later research uses of those cells do not become morally acceptable merely because the research is separated from the derivation. Under this reasoning, the chain of complicity holds even if the actual ES cells used in research or therapy are the product of many later generations or passages of the original immorally derived ES cells.[13]

Such a view of complicity is a reasonable one up to a point. If the original immoral derivation of ES cells occurred with the intent to make later ES cell research possible, then it is reasonable to view the later researchers as complicit in the original derivation, on the causative theory that the derivation would not have occurred if the later use had not been contemplated. Complicity based on causation, however, would not exist if the original immoral derivation of cells would have occurred regardless of the activities of any particular later researcher. Once the ES cells have been derived for particular kinds of research or for particular researchers, they exist and could be used by other researchers. Under a causative theory of complicity, making the ES cells available to later researchers would not make those researchers complicit in the immoral derivation of the ES cells if their research plans or actions had no effect on whether the original immoral derivation occurred.

The later researchers in this case would, of course, be profiting from or making use of that original immoral derivation and, in that sense, benefitting from another person's wrongdoing. However, a "no benefit from another's wrongdoing" theory of complicity seems much too broad to be a guide to moral or social practice. If taken seriously, it would mean that the taint of an original alleged immoral action, no matter how attenuated, could never be removed as long as it were still traceable to the original action.[14] Such a view would make us all morally complicit in any immoral action that at several removes still underlies or contributes to economic and social transactions from which we benefit. It would also bar the use of organs for transplant that resulted from a homicide or suicide, because the recipient would be benefitting from another's immoral action that occurred independently of the recipient's need for an organ transplant.

ETHICS AND POLICY IN RETRIEVAL OF ES CELLS FROM ABORTED FETUSES

Primordial germ cells retrieved from first-trimester aborted fetuses provide one potential source of ES cells for research and therapy.[15] It is unclear at this time whether fetal tissue or human preimplantation embryos will become the main source of ES cells. Current law, however, permits the use of federal funds to derive such cells from the tissue of aborted fetuses, though it prohibits the use of federal funds to isolate them from embryos.

Because primordial germ cells are removed from fetuses after their death, the derivation of ES cells from aborted fetuses does not cause their death as it does in the case of preimplantation embryos. Nor, with a million-plus electively induced abortions occurring annually in the United States, is there a strong basis for the claim that ES cell research using primordial germ cells would cause many women faced with an unwanted pregnancy to have abortions that would not otherwise have occurred simply because of the chance to donate fetal tissue for research.[16]

Under a causative theory of complicity, neither derivation nor later use of ES cells from abortions that would otherwise have occurred would make one morally complicit in the abortion itself because there is no reasonable basis for thinking that donation of tissue for research after the decision to abort has been made would have caused or brought about the abortion. Thus persons who think that induced abortion is immoral could support the use of fetal tissue or ES cells derived from abortions as long as the derivation or later research or therapy had no reasonable prospect of bringing about abortion, just as they could support organ donation from homicide victims without approving of the homicide that made the organs available. To do so, however, such individuals would have to be convinced that research uses of fetal tissue from abortions otherwise occurring would not bring about future abortions or in some way make abortion appear to be a positive, praiseworthy act.

On the other hand, if persons who think that induced abortion is immoral hold the "no benefit from another's wrongdoing" view of complicity, they would object to the derivation of tissue from fetuses that would be aborted regardless of research plans, merely because one would be deriving benefit from what is viewed as an immoral act. It is unclear how many persons opposed to abortion hold the broader, "no benefit" theory of complicity. If they do hold such a view and are consistent, they should also object to transplanting or receiving organs from murder victims because that would constitute benefitting from another's wrongdoing. Similarly, they should object to any benefits that might in some traceable sense also be the result of another's wrongdoing. Because many common activities, practices, and social arrangements may be traceable to some past wrongdoing—e.g., wresting land from Native Americans—persons holding that view, if they are consistent, would have a difficult time living in the contemporary world.

As a policy matter, the question of obtaining tissue for research or therapy from aborted fetuses received enormous scrutiny between 1988 and 1993 after NIH received proposals to use tissue from aborted fetuses to treat Parkinson's disease, diabetes, and other illnesses. The assistant secretary of health declared a moratorium on NIH funding of fetal tissue transplantation research in 1988 and appointed an expert panel—the NIH Fetal Tissue Transplantation Research Panel—to review the issue. The panel concluded that such research could be ethically conducted if the decision to abort were carefully shielded from the decision to donate tissue, thus reflecting a causative view of moral complicity. In recommending NIH funding for such research, the Panel also recommended that no compensation be provided for donation and that no changes in the timing or method of abortion occur to facilitate retrieval of fetal tissue.[17]

Despite these recommendations, Secretary Louis Sullivan chose in 1990 to continue the federal funding moratorium because he doubted that a strict line between the decisions to abort and to donate tissue could be maintained.[18] This decision was applauded by right-to-life groups but was widely criticized in Congress. Twice Congress passed legislation to overturn the funding moratorium, and twice President Bush vetoed the measures, with the vetoes upheld in each case by a narrow margin. On 22 January 1993, newly elected President Clinton used his executive authority to lift the moratorium on federal funding of fetal tissue transplantation research. Congress then codified authority for federal funding of fetal tissue transplantation research by enacting limits akin to those initially recommended by the NIH advisory panel.[19] That law also made it a federal crime to donate fetal tissue for transplantation to a family member or other designated individual.[20]

Under current federal law it is lawful to conduct and federally fund retrieval of ES cells from primordial germ cells removed from fetal tissue donated after an induced or spontaneous abortion as long as no valuable consideration has been paid for the donation, the woman undergoing the abortion gives informed consent to the donation, and there has been no promise to transplant the tissue into a person specified by the donor or to a relative. In addition, the attending physician must provide a statement concerning the woman's informed consent and the method of obtaining the tissue; a statement by the researcher regarding his or her knowledge of the source of the tissue; and an assurance that the researcher has not participated in any decision regarding termination of the pregnancy. NBAC (1999) has recommended that federal funding for retrieval of fetal tissue for stem cell research should continue under these conditions. If these conditions are met, persons who object to abortion but who hold a causative theory of moral complicity have a reasonable basis for finding that retrieval of ES cells from aborted fetuses is ethically and legally acceptable. Since the initial retrieval does not involve causative complicity in the death of the fetus, later research or therapeutic uses of the cells retrieved should be ethically as well as legally acceptable.

ETHICS AND POLICY IN THE RETRIEVAL OF ES CELLS FROM HUMAN PREIMPLANTATION EMBRYOS

A major source of ES cells for research may turn out to be human preimplantation embryos rather than aborted fetuses.[21] As noted above, ES cells derived from embryos are not themselves embryos. However, the need at some point to isolate ES cells for research from live embryos raises ethical concerns of complicity in the destruction of those embryos. Because the embryo is alive at the time that ES cells are removed from the inner cell mass of the blastocyst, research that depends upon the isolation of ES cells from preimplantation embryos raises questions about the permissibility of destroying embryos for research. These questions arise under both the causative and the no-benefit theories of moral complicity, for in either case at least some embryos may be destroyed in order to facilitate or to carry out ES cell research.[22] If the embryos are donated by couples undergoing IVF, the embryos might otherwise have been discarded or kept indefinitely in storage. In some cases they may have been created specifically for the purpose of obtaining ES cells for research. In either case, however, the embryo, unlike the aborted fetus from whom ES cells are retrieved, is alive at the time that the ES cells are removed, even though it may be discarded or never transferred to a uterus if not used in research.

To assess the ethics and policy of ES cell research that is directly destructive of preimplantation embryos, we must revisit debates over embryo research and the status of embryos. Those debates have focused on two main issues: (1) May spare human embryos be destroyed in research when they are donated by couples undergoing IVF treatment for infertility? and (2) May human embryos be created for destructive research without ever having intended to transfer them to the uterus for implantation?

USE OF DONATED SPARE EMBRYOS IN RESEARCH

Normative Concerns

The key ethical and policy debate in embryo research—whether such research should occur at all—has assumed that the embryos used in research would be spare embryos donated by couples undergoing IVF treatment for infertility. Only if such research is deemed acceptable, does the second question—whether embryos may be created for research and then destroyed—arise.

The debate over whether embryos may be destroyed in research when they are donated by couples undergoing IVF has turned on perceptions of the moral status of embryos and the extent to which a rights-based or a deontologic/symbolic view of that status should determine the ethical acceptability of practices involving the first stages of human life. Biologically, there is agreement that the cells of preimplantation human embryos are not yet differentiated into organs or particular tissues. Nor are they clearly committed to individuation because spontaneous twinning can still occur until the time of implantation.[23] Indeed, the embryo's first cellular differentiation is between trophectoderm or placental tissue and the inner cell mass, which contains ES cells. Only after implantation does the embryonic disc and then the primitive streak, from which the brain, the nervous system, and other organs of the body grow, form.

Disagreement, however, attends the moral evaluation of these facts, with differences largely reflecting people's views about abortion. Many people who oppose abortion believe that the fertilized egg's potential to develop into a new human being automatically confers upon it full moral status as a person. They oppose all nontherapeutic embryo research regardless of its benefits, for they view any invasive procedure done to embryos, such as removing stem cells from the inner cell mass, as unjustifiably harming them. Individuals who hold this view are as offended by research with spare IVF embryos as they are with the creation of embryos solely for research. They would oppose any destructive derivation of ES cells from embryos and the research that causes it to occur.

Many other people, however, do not view the previable fetus as an entity that has developed to the point that it has interests that justify overriding a pregnant woman's right to terminate pregnancy. Not surprisingly, persons holding this view about previable fetuses view preimplantation embryos, which are much less developed than fetuses, as too rudimentary in structure or development to have moral status or interests in their own right. For them the attribution of moral status rationally depends upon at least the presence of a nervous system, if not also sentience, and not just its precursor cells. As a result, such individuals maintain that no moral duties are owed to embryos by virtue of their present status and that they are not harmed by research or destruction when no transfer to the uterus is planned. Under this view of embryo status, there is no moral objection to destroying spare, donated embryos to conduct ES cell research. Indeed, if not used in research, those embryos will be discarded or kept indefinitely in frozen storage.

Persons holding the latter view—that the embryo itself lacks interests or rights because of its extremely rudimentary development—do not, however, necessarily view embryos as identical to any other human tissue. Indeed, many such persons would say that embryos, though lacking rights or interests in themselves, deserve "special respect" because of the embryo's potential, if placed in a uterus, to become a fetus and eventually to be born. Even embryos that will not be placed in the uterus have some meaning in this regard for they operate as a symbol of human life or constitute an arena for expressing one's commitment to human life.

This distinction between intrinsic and symbolic valuation of the embryo is at the heart of debate over both abortion and embryo research.[24] One can deny that something has intrinsic value as a moral subject, yet still value it or accord it meaning because of the associations or symbolism that it carries. This distinction has special relevance to the embryo research debate. Even persons who view the embryo as lacking rights or interests in itself are not comfortable with anything at all being done with embryos, for example, using them for toxicology testing of cosmetics or buying and selling them. Accordingly, they would accept the use of spare embryos in research only when there is a good medical or scientific reason for doing so. In effect, the benefits of such research are deemed to outweigh whatever symbolic costs or losses arise from treating an entity that in other circumstances might be transferred to the uterus as if it lacked that potential.

Under this normative approach, embryo research has been deemed acceptable when necessary to pursue a legitimate scientific or medical end that cannot be pursued by other means, when there has been local or national review of the proposed research, and when the embryos have been donated for research with the informed consent of the providing couple. Additional safeguards, such as limiting the research purposes to which embryos might be put or requiring a national ethics advisory board review of such research, would further mark the symbolic importance of spare embryos when they are used in research. The need for additional procedural safeguards depends on the particular political circumstances of the national, state, and institutional context in which they arise. ES stem cell research with spare IVF embryos would clearly meet a high standard of need for conducting research with spare embryos. Its great potential to treat or prevent disease in many persons shows that destructive research with embryos that are unwanted for reproduction will be used for the beneficial purpose of preserving life.

National Advisory Commissions and the Use of Spare Embryos for Research

National advisory commissions in the United Kingdom, Canada, and the United States (as well as in other, although not all, nations) have found embryo research to be ethically acceptable and have recommended government support for it while recognizing the symbolic importance of human embryos, although they have not often specified the normative considerations outlined above. The British acceptance of embryo research was articulated in the landmark 1984 Warnock Report, which found that legally the human preimplantation embryo "did not have the same status as a living child or adult . . ." but ". . . that the preembryo of the human species had a special status."[25] Along with recommendations in many areas for how assisted reproduction should be conducted, the Warnock Report found that research with spare embryos was acceptable for specified purposes. These recommendations laid the foundation for the Human Fertilisation and Embryology Act of 1990, which created a Human Fertilisation and Embryology Authority (HFEA) to license fertility clinics and research projects with human embryos.[26]

The 1990 act, however, prohibits the HFEA from authorizing a research project with embryos "unless it appears to the Authority to be necessary or desirable" to undertake the research for specified purposes. The permissible purposes include treatment of infertility; the acquisition of knowledge about congenital disease, the cause of miscarriage, more effective means of contraception, and preimplantation genetic or chromosomal diagnosis of abnormalities; and "for such other purposes as may be specified by regulations." Under this language, embryos (whether spare or created) could not be used to obtain ES cells to study embryo and tissue development until the Secretary of State for Health issued regulations adding cell and tissue replacement therapies to the list of permitted research purposes.[27]

In the United States, two national ethics review bodies have addressed issues of embryo research prior to the current NBAC attention to the subject.[28] An Ethics Advisory Board (EAB) appointed by the Secretary of Health, Education, and Welfare to advise on federal funding of research involving IVF unanimously agreed in 1979 that "the human preembryo is entitled to profound respect, but this respect does not necessarily encompass the full legal and moral rights attributed to persons."[29] The EAB also found that research with embryos was acceptable for studies related to infertility. Because of then existing DHEW regulations, EAB approval would be required for federally funded embryo research.[30]

When the terms of the EAB members lapsed in the 1980s, neither the Reagan nor Bush administrations appointed new members, thus preventing the requisite EAB approval for federal funding of embryo research. In 1993, the Clinton administration supported a congressional enactment that removed the requirement of EAB review for federally funded embryo research.[31] The director of the National Institutes of Health then appointed a Human Embryo Research Panel (HERP) to recommend guidelines for federal funding of embryo research. The HERP's 125-page report thoroughly canvassed the field and identified many kinds of embryo research, including ES cell research, as acceptable for federal funding with embryos donated by couples undergoing IVF.[32] It also approved the creation of embryos for research when essential to carry out the research. The HERP report clearly rejected the view that embryos have inherent or intrinsic rights, but nevertheless imposed detailed restrictions on their use both to mark their symbolic importance and to show some deference to those who oppose any embryo research. Congress, however, has refused to permit funding of any research with human embryos, regardless of the purpose of the research and source of the embryos.

The promise of ES-based cell or tissue replacement therapies has put the question of federally funded embryo research back on the public agenda, for federal law now prohibits funding of many kinds of ES cell research. At the request of President Clinton, the National Bioethics Advisory Commission has now examined the issue and recommended that "Research involving the derivation and use of ES cells from embryos remaining after infertility treatments should be eligible for federal funding, given an appropriate framework for public oversight and review."[33]

NBAC reached this conclusion after noting that the differences between those who oppose all embryo research and those who support it are largely intractable because "there seems to be no rational way of securing moral agreement in our culture."[34] Instead NBAC sought to develop "policies that demonstrate respect for the alternative points of view and center, where possible, around points of convergence and agreement."[35] It found that there may be a "sufficiently broad, overlapping opinion on the status of embryos to justify, under certain conditions, some research uses of stem cells derived from them."[36] Drawing on Ronald Dworkin's analysis of the acceptance by some abortion opponents of exceptions to a ban on abortion in the case of threat to the mother's life or rape, the NBAC thought that conservatives and liberals might agree on permitting federally funded destructive embryo research "where there is good reason to believe that this destruction is necessary to develop cures for life-threatening or severely debilitating diseases."[37] It also found that the uncertainty about whether such benefits will be forthcoming is outweighed by the great number of potential lives saved if the research leads to successful therapies. Its conclusions and recommendations, however, rest on the assumption that any use of embryos to derive stem cells will occur only when other sources of stem cells for research no longer suffice.

In reaching its conclusion that the ban on federal funding of ES cell research should be partially rescinded, NBAC gave great weight to principles of beneficence and nonmaleficence, which underlay the ethical goals of medicine in conducting research and treating and preventing disease. A ban on federal funding, while permitting private sector research to occur, interferes with such goals and leaves much unregulated embryo research on the fringes of public life. In NBAC's view, "the intentional withholding of federal funds for research that may lead to promising treatments may be considered unjust of unfair."[38]

The overlapping consensus among proponents and some opponents of embryo research described by NBAC assumes limits and regulation in carrying out ES cell research with embryos. Regulatory limits are a key part of that consensus, for they provide accountability, allay public anxiety, and "can ensure greater public accountability and transparency."[39]

To emphasize the sensitivity of such research, NBAC also recommended a national review mechanism for ES cell and other embryo research protocols to "ensure strict adherence to guidelines and standards across the country" and to "provide the public with the assurance that research involving stem cells is being undertaken appropriately."[40] To do this, it recommended establishment of a National Review and Oversight Panel within the Department of Health and Human Services. The Panel would provide "general review of all research protocols in which the use or derivation of human ES cells . . . are considered."[41] Together with a Public Registry of stem cell research proposals, such national ethics review would mark the special importance and meanings attendant on ES cell research and permit public oversight even as it permits such research to occur.

CREATION OF HUMAN EMBRYOS FOR RESEARCH OR THERAPY

Although much ES cell research may occur with cells derived from embryos donated after IVF treatment for infertility, at some point in the future some types of stem cell research, such as efforts to develop or isolate tissue that is compatible with the immune system of prospective patients, may require the creation of embryos. For example, the ability to develop a large library of ES cell genotypes that encode different transplantation antigens so that cell replacement therapies would be available for a large range of the population, likely would depend upon the creation of embryos expressly for that purpose. Or embryos might have to be created by nuclear transfer cloning from a prospective patient's own cells in order to obtain histocompatible ES cells for that patient. The ultimate permissibility of creating embryos solely for ES cell research may thus determine whether many important kinds of ES cell research occur.

The question of the acceptability of the creation of embryos for research has been a heated issue in the embryo research debate, but not between those who oppose and approve of research with spare embryos so much as within the group of those who find destructive research with spare embryos to be morally acceptable. Of course, persons who oppose research with spare embryos on the ground that embryos themselves have intrinsic moral status also oppose the creation of embryos for research. However, persons who approve of research with spare embryos because of their view that embryos lack interests nevertheless disagree about whether embryos should be created for research purposes when there is never any intent to transfer those embryos to the uterus. Thus, a subset of persons exists who approve of research with spare embryos, but object to the creation of embryos solely for research. Presumably these same persons would object to the creation of embryos solely to obtain ES cells for research or therapy. As a result, the ethical controversy over creating embryos specifically for research is likely to surface again if science progresses to the point that creation of embryos becomes necessary to obtain ES cells for certain types of research or therapy.

The opposition to creation of research embryos by persons who approve of research with spare IVF embryos cannot be justified on harm- or rights-based grounds, for those persons also agree that embryos are too rudimentary in development to have interests or rights, and that they are not harmed by destructive research when donated by couples undergoing IVF. Such individuals oppose the creation of research embryos either because of consequentialist concerns about the effect of such practices on other persons or because of deontologic or symbolic/constitutive concerns about showing respect for human life.

The 1994 HERP report mentioned both types of concerns:

> Invoking deeply held and widely shared beliefs about the significance of fertilization as the first step into bringing a potential human being into existence, those opposed to fertilization of oocytes for research argue that this step ought not to be taken solely for research purposes, no matter how important these purposes might be. They maintain that development of embryos expressly for research is *inherently disrespectful of human life,* as well as being open to significant abuses. They also fear that this practice will lead to the *instrumentalization of the preimplantation embryo, and by extension, of other human subjects.* They are particularly concerned that the development of embryos for research *may result in the commodification of embryos and even their commercialization.*[42]

Professors George Annas, Arthur Caplan, and Sherman Elias also have made arguments that combine consequentialist and symbolic deontologic concerns:

> To create embryos solely for research—or to sell them, or to use them in toxicity testing—seems morally wrong because *it seems to cheapen the act of procreation* and turns embryos into *commodities.*

Creating embryos specifically for research also *puts women at risk* as sources of ova for projects that *provide them no benefit*.

The moral problem with making embryos for research is that as a society we do not want to see *embryos treated as products or as mere objects*, for fear that we will *cheapen the value of parenting, risk commercializing procreation, and trivialize the act of procreation*. It is society's moral attitude toward procreation and the interests of those whose gametes are involved in making the embryos that provide the moral force behind the restriction or prohibition of the manufacture of embryos for nonprocreative uses."[43]

Consequentialist Concerns in Creating Research Embryos

The consequentialist arguments against creating research embryos expressed in these excerpts assume that creation of embryos for well-justified research projects will inevitably lead to bad consequences, yet they never show that those consequences are very likely to occur nor the mechanism by which they would come about. For example, Annas, Caplan, and Elias fear that creating embryos for responsible medical or scientific research will quickly lead to the production of human embryos for more trivial uses such as toxicology screening of drugs or cosmetics and the emergence of an industry that buys and sells human embryos. The HERP quotation cites persons who think that creating embryos will "cheapen or demean" respect for other research subjects. Annas, Caplan, and Elias also think that creating embryos for research will cheapen or demean procreation, thereby undermining respect for persons generally. It will also put at risk women who are induced to serve as donors of the eggs from which the embryos will be created.

It is highly unlikely, however, that such effects would occur if embryos retain the special symbolic respect that marks the HERP and NBAC approaches to research with spare embryos. Given the controversial and sensitive nature of creating embryos for research, it is likely, at least in federally funded research institutions, that research embryos will be created only for compelling reasons, for example, when important research cannot be validly conducted with spare embryos. Creation of research embryos for meritorious research is thus unlikely to lead quickly to a wide use of created embryos for less compelling purposes, such as mass toxicology screening. Nor is a market that buys and sells embryos for research or procreative purposes likely to arise. Indeed, leading proponents of creating embryos for research agree that restrictions on the sale of embryos are desirable.[44]

It is similarly unlikely that the creation of embryos for research that could not easily be conducted in other ways will undermine respect for other research subjects. Apparently the theory for such an effect is that by creating human embryos to serve as instruments or means to obtain knowledge, researchers will be more likely to use other human subjects as mere means. This speculation rests on the counterintuitive premise that carefully limited laboratory research involving the creation of research embryos will affect the interactions that *different* clinical researchers have with human subjects in other projects. The claim also ignores the well-entrenched system of IRB review and informed consent that protects the rights and welfare of research subjects. It is highly implausible that acceptable research projects involving created embryos would undermine or weaken this system or decrease respect for human life or persons generally.

The Annas-Caplan-Elias version of the consequentialist argument—namely, that the creation of research embryos will cheapen or demean human reproduction and parenting by commercializing procreation—assumes an unlikely scenario of embryos being created en masse for commercial purposes. Even if such a practice occurred, it still would not follow that all other acts of conception or procreation would be "demeaned," any more than the existence of prostitution (whether legal or illegal) demeans sexual intercourse between spouses or lovers.[45] In any event, it is difficult to see how the creation of embryos for research when there is a good scientific or clinical need for the practice would "cheapen" or "demean" procreation and parenting generally, much less decrease societal respect for human life, nor do the authors specify how such an effect would come about.

A more substantial consequentialist concern is the effect that the practice of embryo creation might have on women who donate the eggs that are fertilized to create embryos for research.[46] Most of the oocytes that would be fertilized to provide embryos for research are likely to be donated by women who are donating eggs for reproductive purposes to an infertile couple or by women who undergoing IVF but who produce more oocytes than they need for themselves. In some cases, however, researchers may have to recruit women to undergo ovarian hyperstimulation and egg retrieval in order to get eggs for research. The

Annas-Caplan-Elias claim that such women would be undergoing significant risks "for no benefit" overlooks the personal benefits those women receive from choosing to serve as contributors to scientific research (a parallel to the altruistic component of oocyte donation for reproductive purposes) and the scientific benefits that the research aims to generate. In any event, all would agree that the rights and welfare of women asked to donate oocytes—whether for reproduction or research—should be carefully protected. They should be fully informed of the risks and benefits of such an act. If they are so informed and nevertheless wish to proceed, there is no justification for permitting egg donation for assisted reproduction but prohibiting it in the case of ES cell research. ES cell research needs are as compelling as the needs of infertile couples.

Deontologic and Symbolic/Constitutive Concerns in Creating Research Embryos

Much of the opposition to the creation of research embryos from persons who reject the notion that embryos have intrinsic moral status rests on deontological or symbolic/constitutive grounds.[47] As the Annas-Caplan-Elias and the HERP quotations show, opponents of creating embryos for research have argued that it is inherently wrong to create an embryo with no intent to transfer it to a uterus because such a practice treats the created embryo as a mere means or instrument to others' ends, thereby expressing "inherent disrespect" for human life. They distinguish the IVF practice of fertilizing more oocytes than can safely be placed in the uterus on the ground that the embryos were created for the purpose of procreation and that *ex ante* fertilization each potential embryo had an equal chance with the others to be transferred to the uterus.[48]

Persons holding the view that the creation of research embryos is wrong try to draw support from the Kantian deontologic tradition that it is wrong to treat human beings as mere means or instruments to the ends of others. This deontologic claim, however, assumes that the preimplantation embryo is already a human being or human subject with interests that are harmed by treatment as a mere means. Yet the persons asserting this ground also agree that spare IVF embryos are too rudimentarily developed to be harmed, and thus may ethically be donated for use in research. Therefore, they cannot object to the creation of research embryos on the ground that doing so would violate ethical norms against using persons as mere means. As they concede, embryos created for research are not persons who can be used to their detriment. If they disagree with this conclusion, then they should object to research with spare embryos as well, for they too become means to serve the needs of others once the decision to use them in research is made.[49]

A more useful way to view nonconsequentialist claims against the creation of embryos for research is as symbolic/constitutive assertions of a person or community's attitudes toward the importance of human life.[50] Because preimplantation embryos are the first stage of a new human life, they ordinarily are created for the purpose of bringing such a life into the world. Accordingly, they function as a powerful symbol of human life and provide the occasion for demonstrating or expressing commitment to human life generally, for example, by condemning creation of embryos for research as being "inherently disrespectful" of human life. In taking such a stance, persons define or constitute themselves as highly protective of human life.

This articulation of the issue tracks very closely the earlier discussion about intrinsic versus symbolic concerns in research with spare embryos and points to a similar resolution. Although embryos do not themselves have rights, they are an occasion for expressing or symbolizing one's views about the importance or value of human life, thereby constituting one's moral or national character in the process. People differ, however, over the degree and the intensity of the symbolic associations that attach to non-rights-bearing entities such as embryos. The importance of signifying or constituting a highly protective attitude toward human life by objecting to certain kinds of embryo research is thus more determined by personal or public policy preferences than it is by the obligations of moral duty.

Individuals might thus accept that the embryo has symbolic/constitutive importance, but find that in particular circumstances other actions connected with the protection of human life also have importance.[51] As we have seen, one could reasonably find that the research and clinical benefits from research with spare IVF embryos outweighs the symbolic detriment that might accrue from using discarded spare embryos as a means of producing knowledge. A similar judgment could be made about the creation of embryos for research. If one concluded that the benefits of the research made possible by the creation of embryos outweighed the symbolic or expressive detriments that some persons perceive to flow from such a practice, then one would find the practice to be ethically acceptable. A good example would be research that creates embryos through nuclear transfer cloning to obtain recipient-compatible tissue or cells for transplant.

In making such judgments, one inevitably compares the symbolic or constitutive costs of a particular research practice with its benefits—e.g., does creating research embryos when necessary to conduct important research so diminish respect for human life that those research benefits should be foregone in order to demonstrate respect for human life in this way? Given the highly personal nature of symbolic/constitutive claims, there is no definitive answer to this question. The answer will depend on the nature and purpose of the research, the availability of alternatives, a person's values and commitments, and public perceptions of a practice.

The ethical acceptability of creating embryos for ES cell research will thus turn on the symbolic/constitutive meanings associated with such a practice in light of the benefits that such research will provide. The conclusion will depend on individual perceptions of the importance of the ES cell research project and the harm to respect for human life that creation of embryos for ES cell research is perceived to cause. Some persons would argue that creation of research embryos raises symbolic or expressive harm beyond that which exists with research with spare embryos and which cannot be justified by the uncertain benefits that the research in question might bring. Other persons, however, might reasonably conclude that the additional symbolic harm of creating embryos for ES cell research that could not otherwise validly occur is minimal. To mark the greater symbolic importance of such research, they might limit it to the most important kinds of research and require additional review procedures.

National Advisory Commissions on Creating Embryos for Research

The United Kingdom distinguishes acceptable embryo research in terms of the purpose of the research, not the source or mode of creation of the embryos used in the research. This position derives from the 1984 Warnock Committee, which by a nine to seven vote found that embryos could be created solely for research.[52] In 1990 Parliament, in enacting legislation implementing the Warnock Committee recommendations, rejected by a large majority a prohibition on the creation of embryos for research. The Human Fertilisation and Embryology Authority has since licensed several projects involving the creation of embryos for research related to improving infertility treatments. However, until regulations are issued to allow ES cell research with embryos, no embryos, whether spare or created for research, may be used for ES cell research in the United Kingdom.

In the United States, although the 1979 Ethics Advisory Board did not address the question of creating embryos for research, the Human Embryo Research Panel thoroughly addressed the issue in 1994. The Panel recommended, by a large majority, federal funding for creating embryos for research when the research could not otherwise validly occur:

> The Panel recognizes . . . that the preimplantation embryo merits respect as a developing form of human life and should be used in research only for the most serious and compelling reasons. There is also a possibility that if researchers have broad permission to develop embryos for research, more embryos may be created than justified. In order to minimize this, the Panel believes that the use of oocytes expressly for research should be allowed only under the following two conditions:
>
> —When the research by its very nature cannot otherwise be validly conducted. Examples of studies that might meet this condition include oocyte maturation or oocyte freezing followed by fertilization and examination for subsequent developmental viability and chromosomal normalcy and investigations into the process of fertilization itself (including the efficacy of new contraceptives).
>
> —When the fertilization of oocytes is necessary for the validity of a study that is potentially of outstanding scientific and therapeutic value.[53]

The HERP also specifically addressed the creation of embryos for ES cell research. Although the Panel deemed such research not to be acceptable for federal funding in 1994, it found that such research warranted further review in the future. In so recommending, the Panel noted the potential utility of ES cells for transplantation and tissue repair, and circumstances in which creation of embryos to provide ES cells for research might be warranted to create a large library of ES cell genotypes to serve the cell or tissue replacement needs of a large range of the population. It thus implied that the presumption against federal funding to create embryos solely for research might be overcome when a clear research need could be demonstrated.

The Panel also noted that future developments might reduce the need to fertilize oocytes solely for research. For example, the ability to alter ES cell genes to control transplantation antigens would obviate the need to create embryos to obtain a range of lineages compatible with many different genotypes. Similarly, development of the ability to "transfer nuclei from differentiated adult cells into enucleated eggs and obtain normal development to the blastocyst stage" would provide a genome identical to that of the adult donor but would not involve the fertilization of male and female gametes.[54]

The HERP's carefully nuanced recommendations for federal funding of meritorious research that required the creation of research embryos did not fare well in the political process. On the day the Panel's recommendations were announced, President Clinton stated that he opposed federal funding of research embryos regardless of the need or purpose of the research.[55] Congress then banned federal funding of all embryo research, regardless of the purpose or source of embryos.[56] That ban has been reenacted every year since then.

NBAC's report on embryonic stem cell research also discusses the creation of research embryos for ES cell research. Its discussion, however, focuses more on the deontologic or symbolic opposition to such a practice than on consequentialist concerns. It approaches the issue in terms of differences between creating embryos for reproduction and creating them for research. The latter concern appeals to "arguments about respecting human dignity by avoiding instrumental use of human embryos: using embryos merely as a means to some other goal treats them without respect or concern."[57] Such a concern, as I have argued, is, for individuals who do not oppose research on spare embryos, clearly symbolic or expressive, for there can be no obligation to treat a thing as a moral end in itself if it never will have the intrinsic characteristics of a moral entity, which is the case with embryos created for ES cell research.

NBAC, however, did not have to take a position on the issue, for it found that:

> At this time, there are no persuasive reasons to provide federal funds for the purpose of making embryos via IVF solely for the generation of ES cells. More research should be done on stem cells derived from aborted fetuses and embryos remaining after infertility treatment to determine the extent of the need of these additional sources of embryos for research.[58]

If the science already had progressed to the point that there were a strong medical or scientific need to create embryos for ES cell research (because spare embryos were not available for donation or because person-specific embryos were needed), NBAC would have conceivably taken a different position. The unarticulated symbolic/expressive normative approach that underlies the NBAC position on ES cell research with spare embryos would, as a similar position did for the HERP, permit approval of the creation of embryos for ES cell research when necessary to carry out important research, such as the study of embryos created by cloning to provide histocompatible tissue for transplant.

PUBLIC RESEARCH POLICY AT THE BEGINNING OF LIFE

With advances now rapidly occurring in the understanding and manipulation of human ES cells, current federal policy against funding any destructive embryo research needs to be reassessed to avoid discouraging or preventing such research. The Raab distinction between federal funding of use rather than derivation of ES cells may have some political appeal[59] because it would permit some federally funded ES cell research to proceed without necessitating a change in the present law, as federal support of ES cell derivation would require. A ban, however, on federal funding of either the derivation or the use of ES cells creates a barrier to further progress in ES technology. For example, permitting federally funded research with ES cells derived in the private sector still leaves researchers facing intellectual property and other barriers to obtaining privately derived ES cells. Those barriers could deter some researchers from undertaking ES cell research and thus slow the pace of progress in the field.

Only the group of persons who view fertilized eggs, early embryos, and fetuses as themselves persons or subjects with intrinsic rights strongly object to the use of aborted fetuses or spare embryos for ES cell research.[60] Their view of the intrinsic moral status of embryos—that spare or created embryos are persons that cannot be destroyed in research—is not required by our legal and constitutional traditions and should not drive federal research policy. Indeed, the basis in this position for imputing intrinsic value to the embryo is more the reflection of religious or spiritual perspectives than it is of a persuasive philosophical or normative position about when protected human life begins. It has been rejected by most advisory commissions that have studied the matter.

Precisely because people differ so deeply over personal spiritual and value commitments, one group should not erect its own view of the matter into public policy. Indeed, doing so exacts a high cost from those who would benefit from such research. Their health needs are held hostage to a set of religious positions held by a minority that has inordinate sway in the legislative process. Prohibition of federal funding for ES cell research is also a major barrier to further research and development in this field. In addition, it leaves private sector researchers without a clear set of ethical or regulatory guidelines for the ES cell research, which they do, and will continue to, conduct.

A more fruitful approach to research issues at the beginning of life is to recognize that for most persons the ethical or normative questions that arise are less about the duties intrinsically owed to embryos or fetuses than they are about symbolizing or expressing the high respect that most persons have for human life generally. This respect is shown, not by banning all research with embryos or aborted fetuses, but by allowing such research only when good reasons exist for engaging in it and an institutional, or even national, review process to assess those reasons has been implemented.

National review bodies in the United Kingdom and the United States that have considered embryo research have, with some variation in the details of justification, taken such a normative approach to issues of embryo research. They have consistently refused to find that embryos are themselves persons with intrinsic rights, and they have recommended that research be permitted with spare embryos when necessary for good medical or scientific purposes. The Warnock Committee and the NIH Human Embryo Research Panel also have supported the creation of embryos for research when important research could not otherwise occur, including ES cell research when that is shown to be necessary.

NBAC has re-analyzed these questions in the context of ES cell research and has come to similar conclusions. Congress should enact NBAC's recommendations and allow federal support for both the derivation and the use of ES cells from spare embryos to occur under a system of national review. If ES science progresses to the point that researchers have to create embryos with particular genomes in order to overcome immunological incompatibility in prospective patients, then Congress should also provide federal funds for that research.[61] Embryo research policies, when structured and regulated to mark the special importance of this category of research, can reinforce respect for the earliest stages of human life even as material from those stages is used to help others.

NOTES

1. James A. Thomson et al., "Embryonic Stem Cell Lines Derived from Human Blastocysts," *Science* 282 (1998): 1145–47; Michael J. Shamblott et al., "Derivation of Pluripotent Stem Cells from Cultured Human Primordial Germ Cell," *Proceedings of the National Academy of Sciences USA* 95 (1998): 13726–31.

2. Eliot Marshall, "A Versatile Cell Line Raises Scientific Hopes, Legal Questions," *Science* 282 (1998): 1014–15.

3. Gretchen Vogel, "Harnessing the Power of Stem Cells," *Science* 283 (1999): 1432.

4. Public Law 105–277, H.R. 4328, sect. 511 (October 21, 1998).

5. Eliot Marshall, "Varmus Grilled over Breach of Embryo Research Ban," *Science* 276 (1997): 1963.

6. Harriet Raab, *Federal Funding for Research Involving Human Pluripotent Stem Cells*, memorandum to Harold Varmus, Director, NIH, January 15, 1999.

7. Ibid., p. 2. Although the term "any organism" could refer to non-human organisms as well, it is clear from the context that the statute in question is meant to apply to human organisms.

8. Ibid., pp. 2–3.

9. One research team has reported that mouse ES cells, when combined with trophectoderm cells from another mouse embryo, could pro-

duce a new individual, thereby undercutting the claim that pluripotent ES cells are not totipotent in that they could not produce the birth of another individual. (See Andras et al., "Derivation of Completely Cell Culture–Derived Mice from Early-Passage Embryonic Stem Cells," *Proceedings of the National Academy of Sciences USA*, 90 [1993]: 8424–28.) If this claim is true, it would undercut the biologic basis for HHS Counsel Raab's legal opinion that pluripotent cells are not embryos within the meaning of the current federal ban on federal funding of embryo research. The claim would not, however, prevent the use of such cells in research or therapy if one accepted that totipotent cells or embryos could ethically be used in destructive research. The author thanks Andrea Bonnicksen for helpful comments on an earlier draft and Lee Silver for bringing the study to my attention.

10. Nicholas Wade, "Ruling in Favor of Stem Cell Research Draws Fire of Seventy Lawmakers," *New York Times*, February, 17, A12.

11. Matthew Davis, "Presidential Bioethics Advisors May Come Down on the Side of Federal Funding of Research on Stem Cells Derived from Embryos," *Washington Fax*, March, 4, 1999, 1–2.

12. Matthew Davis, "Administration Supports Publically Funded Stem Cell Research, but Not to the Point of Seeking Changes in the Embryo Research Ban," *Washington Fax*, July 19, 1999, 1–3.

13. Many persons, of course, do not think that the original derivation of ES cells either from aborted fetuses or the abortions that make such donation possible or from live embryos is immoral, and they thus do not see the problem of moral complicity that those who object to ES cell removal do.

14. For example, the United States Supreme Court's equal protection jurisprudence has rejected a "benefit from" view of past racial and gender discrimination in assessing race or gender-neutral public policies that have a disproportionate or disparate racial or gender impact. See *Massachusetts v. Feeny* (442 U.S. 256 [1979]), in which past discrimination against women by the military does not render state preferences for veterans for civil service jobs discriminatory under the fourteenth amendment.

15. Shamblott et al., "Derivation of Pluripotent Stem Cells."

16. National Bioethics Advisory Commission (NBAC), *The Ethical Use of Human Stem Cells in Research*, draft report, July 9, 1999, III–4.

17. National Institutes of Health (NIH), *Report of the Panel on Fetal Tissue Transplantation Research* (Bethesda, Md.: NIH, 1989). The Polkinghome Committee, which advised the United Kingdom government on fetal tissue transplantation, made similar recommendations. (See U.K., *Review of the Guidance on the Research Use of Fetuses and Fetal Material*, Department of Health and Social Security, 1989).

18. John A. Robertson, *Children of Choice: Freedom and the New Reproductive Technologies* (Princeton, N.J.: Princeton University Press, 1994), p. 209.

19. *NIH Health Revitalization Act of 1993*, Public Law 103–43, sect. 111, 107.

20. John A. Robertson, "Abortion to Obtain Fetal Tissue for Transplant," *Suffolk University Law Review* 27 (1993): 1362–69.

21. Thomson et al., "Embryonic Stem Cell Lines."

22. Persons who hold a causative theory of complicity and who believe that embryo destruction is immoral should object to ES cell research only when it can reasonably be shown that embryos would not have been destroyed if the ES cell research had not been planned or contemplated. However, those same persons need not object to ES cell research that cannot reasonably be shown to have brought about the destruction of the embryos.

23. Clifford Grobstein, "The Early Development of Human Embryos," *Journal of Medicine and Philosophy* 10 (1985): 213–20.

24. Robertson, *Children of Choice*. Ronald Dworkin makes a similar distinction in his discussion of abortion in *Life's Dominion* (New York: Alfred A. Knopf, 1993) with his use of the labels "derivative" and "detached" to correspond to the notions of "intrinsic" or "rights-based" and "symbolic," "expressive," or "constitutive" protection used here. John Fletcher ("Deliberately Incrementally on Human Pluripotential Stem Cell Research," NBAC background papers on embryonic stem cell research, 1999), in an important background paper prepared for NBAC, has skillfully shown how Dworkin's distinction helps cross the gap to those who highly value early embryos and fetuses. His overall analysis is similar to the analysis contained in this article.

25. U.K., Report of the Committee of Inquiry into Human Fertilization and Embryology, Department of Health and Social Security, 1984.

26. U.K., Human Fertilization and Embryology Act, 1990.

27. Eliot Marshall, "Britain Urged to Expand Embryo Studies," *Science* 282 (1999): 2167–68.

28. Some state laws, which resulted from state efforts in the mid-1970s to block fetal research in the wake of *Roe v. Wade*'s 1973 constitutional recognition of a right to abortion, make research with human embryos for any purpose and from any source a crime. (See John A. Robertson, "Embryo Research," *Western Ontario Law Review* 24 (1986): 15–37 and Lori Andrews, "State Regulation of Embryo Research," in *Papers Commissioned for the Human Embryo Research Panel* (Bethesda, Md.: National Institutes of Health, 1994). Federal courts have struck down bans on embryo research in three states (Louisiana, Illinois, and Utah) on grounds of vagueness and interference with procreative rights (*Margaret S. v. Edwards*, 794 F.2d 994 [5th Cir. 1986]; *Lifchez v. Hartigan*, 735 F. Supp. 1361 [N.D. Ill], affd 914 F.2d 260 [7th Cir. 1990], cert. denied sub nom., *Scholberg v. Lifchez*, 498 U.S. 1069 [1991]; *Jane L. v. Bangerter*, 61 F.3d 1493 [10th Cir. 1995]). Laws in the states that now ban embryo research will remain a significant barrier to ES cell

research regardless of the source of research funding, the purpose of the research, or the source of embryos. However, those laws would not ban research with ES cells once they have been isolated or cultured, for they would then no longer fit the definition of embryo or fetus contained in those statutes.

29. U.S. Department of Health, Education, and Welfare (DHEW), *HEW Support of Reseach Involving Human In-Vitro Fertilization and Embryo Transfer*, 1979, p. 44; Ethics Advisory Board, "Report and Conclusions," *Federal Register* 44 (1979): 35.033–58.

30. "Research Involving Pregnant Women of Fetuses," *Federal Register* 45 (2001): 46.204(d).

31. Public Law 103–43, 121 (c) (1993).

32. NIH, *Report of the Human Embryo Research Panel*, 1994.

33. NBAC, *The Ethical Use of Human Stem Cells*, p. IV–9.

34. Ibid., p. IV–14, quoting Alastair McIntyre.

35. Ibid., p. IV–16.

36. Ibid.

37. Ibid., p. IV–18.

38. Ibid., p. V–9.

39. Ibid., p. VI–17.

40. Ibid., p. VI–20.

41. Ibid., p. V–22.

42. NIH, *Report of the Human Embryo Research Panel*, p. 42, emphasis added.

43. George Annas, Arthur Caplan, and Sherman Elias, "The Politics of Human-Embryo Research—Avoiding Ethical Gridlock," *New England Journal of Medicine* 334 (1996): 1331.

44. NIH, *Report of the Human Embryo Research Panel.*

45. Margaret Jane Radin, *Contested Commodities* (Cambridge: Harvard University Press, 1996); Scott Altman, "(Com)modifying Experience," *Southern California Law Review* 65 (1991): 293–340.

46. Nicole Gerrand, "Creating Embryos for Research," *Journal of Applied Philosophy* 10 (1993): 175–87.

47. In Dworkin's terminology, such concerns are "detached" from the interests of actual persons and not "derivative" of their status as rights-holders.

48. NIH, *Report of the Human Embryo Research Panel;* Annas, Caplan, and Elias, "The Politics of Human-Embryo Research."

49. Gerrand, "Creating Embryos for Research."

50. John A. Robertson, "Symbolic Issues in Embryo Research," *Hastings Center Report* 25, no. 1 (1995): 37–38.

51. Such a cost-benefit judgment explains the acceptability in IVF therapy of fertilizing more oocytes than can be safely transferred to the

uterus, with the excess then discarded or never transferred. Although this practice causes embryos to be created and then destroyed in order to achieve pregnancy and childbirth, it serves a legitimate, valued need that negates the disrespect for human life that symbolically might occur if embryos were created and then destroyed for trivial purposes.

52. U.K., *Report of the Committee of Inquiry into Human Fertilization and Embryology*.

53. NIH, *Report of the Human Embryo Research Panel*, pp. 44–45.

54. Ibid., p. 79.

55. William Jefferson Clinton, "Statement by the President on NIH Recommendation Regarding Human Embryo Research," *U.S. Newswire*, December 2, 1994.

56. Public Law 104–99, sect. 128 (January 26, 1996).

57. NBAC, *The Ethical Use of Stem Cells in Research*, p. V–10.

58. Ibid., p. V–10.

59. Davis, "Administration Supports Publically Funded Stem Cell Research."

60. If such individuals hold a causative view of complicity, they should accept derivation of primordial germ cells from aborted fetuses when the derivation is clearly separated from the decision to abort, as now exists in federal law. They should also accept ES cell research with cells destructively derived from embryos as long as the research was not a cause of the derivation.

61. If creation of embryos is necessary to obtain ES cells of particular genotypes, as might occur to create a universal library of replacement tissue or to tailor replacement tissue to the recipient via nuclear transfer cloning, then fertilization of oocytes or creation of embryos through cloning would be "necessary for the validity of the study," and should be permitted. (See John A. Robertson, "Two Models of Human Cloning," *Hofstra Law Review* 27 [1999]: 609–39.)

14
The Ethics and Politics of Small Sacrifices in Stem Cell Research

Glenn McGee and Arthur Caplan

Pluripotent human stem cell research may offer new treatments for hundreds of diseases,[1] but opponents of such research argue that pluripotent stem cell therapy comes attached to a Faustian bargain: the destruction of many frozen embryos for every new cure. The National Bioethics Advisory Commission (NBAC), the Geron Ethics Advisory Board (GEAB), and many scholars of bioethics, including, in these pages, John Robertson, have raised interesting ethical issues about potential stem cell research.[2] But nothing holds the attention of the public or lawmakers like the brewing battle between millions of ill Americans who favor stem cell research and millions who oppose the destruction of any fetus or embryo for any purpose. The plan to sacrifice embryos for a revolutionary new kind of research has reawakened a long-dormant academic debate about the morality of destroying developing human life.

Glenn McGee and Arthur Caplan, "The Ethics and Politics of Small Sacrifices in Stem Cell Research," *Kennedy Institute of Ethics Journal* 9, no. 2 (1999): 151–58.

In these pages, Robertson (like the GEAB and NBAC) argues that one's position on whether pluripotent stem cells can be used in research depends on one's view of the intrinsic and symbolic status of the embryo.[3] In essence, Robertson and many others argue that one's position on stem cell research pretty much boils down to one's position on abortion. But that is too simple a framework.

Robertson and NBAC make two mistakes. First, they duck the task of identifying and analyzing criteria for justifying the sacrifice of human cells and human life in the name of research. All who write about pluripotent stem cells refer to respect for pro-life views about symbolic moral status, and GEAB in particular claims that symbolic status is the reason such research should be therapeutic in nature. Yet so far no one has explained what kind of destruction is appropriate and of what sort of creatures and for what specific goals and contexts.

Second, Robertson, NBAC, and GEAB take what we call an "accommodationist" posture toward political opposition to stem cell research. In their haste to avoid rehashing the abortion debate, many proponents of stem cell research, including the National Institutes of Health Counsel, Harriet Raab,[4] cede too much ground to foes of abortion procedures, allowing too many arguments to go unanswered. They thus miss critical opportunities to engage in ethical debate about what sorts of rights and duties attend the making and management of embryonic life.

MORAL SACRIFICE

It seems to us that the central moral issues in stem cell research have less to do with abortion than with the criteria for moral sacrifices of human life. Those who inveigh against the derivation and use of pluripotent stem cells make the assumption that an embryo has not only the moral status of human person, but also a sort of super status that outweighs the needs of others in the human community. It is wrong to abort or kill a human, they argue, and thus it is wrong to kill an embryo. But this argument, which is problematic when made about abortion more generally, is doubly so when made against the derivation of pluripotent stem cells from embryos.

Even if frozen human embryos are persons, symbolically or

intrinsically, this in no way entails the right of a frozen embryo to gestation, or to a risk-free pathway into maturation. Adult and child human beings' right "to life" is, considered constitutionally and as a moral problem, at best a negative right against unwarranted violence by the state or individuals. There are, sadly, few positive rights involved. Americans cannot, for example, claim a right against the state to protect them against disease, disasters, adverse weather, and other acts of nature. If a frozen human embryo is a full human person, it still has no right to life per se, but rather a negative right against unwarranted violence and a weak positive right to a set of basic social services (police protection, fire protection, and the like). The question remains as to what constitutes unwarranted violence against an embryo, and for what reasons might an embryo ethically be destroyed—e.g., in the interest of saving the community. Adults and even children are sometimes forced to give life, but only in the defense or at least interest of the community's highest ideals and most pressing interests. One would expect that the destruction of embryonic life, whatever its moral status, would also take place only under the most scrupulous conditions and for the best communal reasons. It bears noting that only those who consistently oppose all violence, destruction, or killing of any kind in the name of the state, the church, or the community can rationally oppose the destruction of an embryo solely by virtue of its status as a human person.

It remains to be shown what the common good is, and what sort of sacrifice an embryo should make in its interest. It is commonly held that no human being should be allowed to lie unaided in preventable pain and suffering. The desire to ameliorate the suffering of the ill motivated Hippocrates, St. Francis of Assisi, Cicero, and Florence Nightingale. It is a central tenet of contemporary medicine that disease is almost always to be attended to and treated because it brings such pain and suffering to its victims and to their family and communities. Trade-offs are made in the treatment of disease, against cost and other competing social demands. But both the Western ethic of rescue and the practical structure of contemporary health care and other social institutions make it clear that among the deepest moral habits of human life is that of compassion for the sick and vulnerable. One of the compelling tenets of the movement to prevent abortion is the argument that a pregnancy ought not be terminated for superficial reasons, but should be viewed as a responsibility to aid the developing human life and to prevent it from needless suffering.

It is the moral imperative of compassion that compels stem cell research. Stem cell research consortium Patient's CURe estimates that as many as 128 million Americans suffer from diseases that might respond to pluripotent stem cell therapies. Even if that is an optimistic number, many clinical researchers and cell biologists hold that stem cell therapies will be critical in treating cancer, heart disease, and degenerative diseases of aging such as Parkinson's disease. More than half of the world's population will suffer at some point in life with one of these three conditions, and more humans die every year from cancer than were killed in both the Kosovo and Vietnam conflicts. Stem cell research is a pursuit of known and important moral goods.

WHAT IS DESTROYED?

The sacrifice of frozen embryos is a curious matter. Set aside, again, the question of whether a frozen embryo is a human life or human person. Grant for a moment that a one hundred–cell human blastocyst, approximately the size of the tip of an eyelash and totally lacking in cellular differentiation, is a fully human person. What does such a person's identity mean, and in what ways can it be destroyed? What would it mean for such a person to die? When could such a death be justified? These questions require a new kind of analysis.

The human embryo from which stem cells are to be taken is an undifferentiated embryo. It contains mitochondria, cytoplasm, and the DNA of mother and father within an egg wall (which also contains some RNA). None of the identity of that embryo is wrapped up in its memory of its origins: it has no brain cells to think, no muscle cells to exercise, no habits. The one hundred–cell embryo has one interesting and redeeming feature, which as best anyone can tell is the only thing unique about it: its recombined DNA. The DNA of the embryo contains the instructions of germ cells from father and mother, and the earliest moments of its conception determined how the DNA of mother and father would be uniquely combined into a human person. The DNA of that person will, if the embryo survives implantation, gestation, and birth, continue to direct many facets of the growth and identity of our human person. At one hundred cells, nuclear DNA is the only feature of the embryo that is not replaceable by donor components without compromising the critical features of the initial recombination of maternal and paternal genetic material after sex (or, in this case, in vitro).

Opponents of stem cell research make an antiabortion argument, namely that the harvesting of pluripotent stem cells will require the destruction of the embryo. But while the cytoplasm, egg wall, and mitochondria of the embryo are destroyed, we just noted that none of these cellular components identifies the embryo at the one hundred–cell stage. The personifying feature of a one hundred–cell blastocyst is its DNA. Pluripotent stem cells from the harvested embryo are directed to form cell lines, each cell of which contains, in dormant form, the full component of embryonic DNA. The DNA in the cell lines has a much greater chance of continuing to exist through many years than does the DNA of a frozen embryo (which in most cases already will have been slated for destruction by the IVF clinic that facilitated the donation, and which would have no better than a 5 to 10 percent chance of successful implantation in any event). Although most Americans are opposed to the "cloning" of adult human beings, it might be possible to harvest DNA from any of the stem cell–based cell lines to make a new, nuclear transfer–derived embryo, or in fact to make five or ten embryos, each of which would possess all of the DNA of the original embryo. In this sense, the critical, identifying features of the embryo would never have been destroyed in the first place, unless what one means by "destroying" an embryo is the loss of its first egg wall, cytoplasm, and mitochondria. The transfer of the nucleus from an embryo to an enucleated egg is a bit like a transplant, though here the donor and the donation are both the DNA. In the case of embryos already slated to be discarded after IVF, the use of stem cells may actually lend permanence to the embryo. Our point here is that the sacrifice of an early embryo, whether it involves a human person or not, is not the same as the sacrifice of an adult because the life of a one hundred–cell embryo is contained in its cells' nuclear material.

The task of balancing sacrifice in the community is one encountered by Solomon in the Judeo-Christian religious texts. Our institutions must enable us in the community to debate and identify the ideals that merit sacrifice, and the loss must be weighed with justice in mind. An embryo cannot reason and it cannot reject a sacrifice or get up and leave the community. For those who feel special responsibility to embryos, the vulnerability of the frozen embryo may suggest special consideration of the kind given to all moral actors in society who are for one reason or another without voice. The question remains, though: What need is so great that it rises to the level where every member of the human family, even the smallest of humans might sacrifice? Already it is clear that we believe that no need is more obvious or compelling than the suffering of half the world at the hand of miserable disease. Not even the most insidious dictator could dream up a chemical war campaign as horrific as the devastation wrought by Parkinson's disease, which destroys our grandparents, parents, and finally many of us.

If Parkinson's disease, or for that matter if a dictator could only be stopped through the destruction of an infant, most every human would blanch at the idea of such sacrifice. But that is not what is asked here. Even those who hold that an embryo is a person will not want to argue that the life of a one hundred–cell embryo is contained in its inessential components. Assuming that a developing embryo can be salvaged by transplanting its DNA, as we have described, it seems unreasonable to oppose the destruction of the embryo's external cellular material or to fear that the one hundred–cell embryo is killed in the transfer. The identity of the human embryo person, if it is a symbolic or intrinsic person, is tied at that stage to the DNA. That DNA is not lost or even injured by the harvesting of embryonic stem cells. This is not the sacrifice of the smallest and most vulnerable among us. We are debating the potential for temporary transplant of undifferentiated tissue and the DNA of such a "person," rather than the imminent destruction of an embryo discarded by a clinic. It is difficult to imagine those who favor just war opposing a war against such suffering given the meager loss of a few cellular components.

POLITICAL ACCOMMODATION

The road to a democratic debate about stem cell research is difficult, as Robertson acknowledges. We entirely agree with Robertson that NIH, and researchers working on stem cell–based cell lines, need not be considered complicit in the destruction of an embryo far removed in time and space.[5] However, complicity assumes someone has done something wrong. Critics and NBAC are thus right to call attention to the amoral nature of the Raab opinion. It could not be more plain that NIH (and some politicians who favor stem cell research) initially hoped to sidestep entirely any ethical debate over the use of cell lines derived from the destruction of embryos. As we have argued above, it hardly does justice to those who believe that embryos are human persons to sacrifice those persons in a cloud of deception and cowardice. Moral sacrifice demands the highest accountability of our social institutions and especially those entrusted with writing and litigating the theory and policy that guides bodies like NIH.

Since at least 1980, the market on "family values" has been cornered by one side of the political spectrum. Anticipating that the public prefers platitudes about abortion to complex positions, most in American political office do not even discuss abortion and embryo research beyond a simple reference to their pro- or anti-abortion voting record. Meanwhile, the last twenty years have seen conservatives develop extraordinarily rigid and arcane positions on virtually every area of "family values."

Again and again, bioethicists' and politicians' policy of appeasing conservative views on family values has resulted in bad policy and balkanized public debate. Embryo research and fetal tissue are particularly instructive in this regard. Without an engagement of the moral questions involved and without an explicit moral framework to explain the decisions, the nation found itself with bans on embryo research and fetal tissue transplantation research that the majority of citizens did not support.

NIH Counsel Raab, National Institutes of Health Director Harold Varmus, NBAC, the GEAB, and in these pages Robertson each give too much moral ground by ceding the illogical (embryos are special people who can never be allowed to die) and bizzare (researchers must not be complicit in the death of an embryo) arguments made by opponents of stem cell research. The sight of scientists rushing to find a legal opinion that legitimizes their work is not a pretty one, nor is that of politicians running scared from those who threaten to shut down the National Institutes of Health in the interest of preserving frozen embryos in stasis. We must be conscious of the role of politics,[6] but the law and the facts do not help with stem cell research. The issues here are novel and they are hard, but mostly they require philosophical innovation about what an embryo is and how we are to treat embryonic material in a time of stem cell research. Our argument here is that no embryo need be sacrificed, but we must alter the terms and goals of our debate to frame an appropriate moral framework for dealing with embryos.

NOTES

1. James A. Thomson et al., "Embryonic Stem Cell Lines Derived from Human Blastocysts," *Science* (1998): 1145–47; Gretchen Vogel, "Harnessing the Power of Stem Cells," *Science* 283 (1999): 1432.

2. National Bioethics Advisory Commission (NBAC), *The Ethical Use of Human Stem Cells in Research*, draft report, July 9, 1999; Geron Ethics Advisory Board (GEAB), "Research with Human Embryonic Stem Cells: Ethical Considerations," *Hastings Center Report* 29, no. 2 (1999): 36–38; Glenn McGee and Arthur Caplan, "What's in the Dish? [Symposium on ethical issues in stem cell research]," *Hastings Center Report* 29, no. 2 (1999): 38–41; John Fletcher, "Deliberately Incrementally on Human Pluripotential Stem Cell Research," NBAC background papers on embryonic stem cell research, 1999; John A. Robertson, "Ethics and Policy in Embryonic Stem Cell Research," *Kennedy Institute of Ethics Journal* 9 (1999): 109–36.

3. Robertson, "Ethics and Policy in Embryonic Stem Cell Research."

4. Harriet Raab, *Federal Funding for Research Involving Human Pluripotent Stem Cells*, memorandum to Harold Varmus, Director, NIH, January 15, 1999.

5. Robertson, "Ethics and Policy in Embryonic Stem Cell Research."

6. Glenn McGee, "Pragmatic Method in Bioethics" in *Pragmatic Bioethics*, ed. Glenn McGee (Nashville, Tenn.: Vanderbilt University Press, 1999), pp. 31–73.

15
The Ethical Case against Stem Cell Research

Søren Holm

INTRODUCTION

The possibility of creating human embryonic stem cell lines from the inner cell mass of blastocysts has led to considerable debate about how these scientific developments should be regulated. Part of this debate has focused on the ethical analysis and part on how this analysis should influence policymaking.[1]

In this [chapter], I want to look at both issues and present what I believe are the best arguments against the derivation of human stem cells in a way that leads to the destruction or killing of the human entity of which they are a part. My discussion will thus cover:

human embryonic stem cells derived by killing the embryo
human fetal stem cells derived by killing the fetus
human infant stem cells derived by killing the infant
human adult stem cells derived by killing the adult

Reprinted from *Cambridge Quarterly of Healthcare Ethics* 12 (2003): 372–83.

The discussion falls into three parts. I first present a critical analysis of the most common arguments showing that the derivation and use of stem cells are morally unproblematic (pro–stem cell arguments). I then look at the arguments showing that the derivation and use of stem cells are problematic (anti–stem cell arguments). Following this, I analyze the policy stalemate that these two opposing views lead to and consider the suggestion that the stalemate can be resolved by requiring consistency between areas of regulation.

THE REDUCTIO OF THE PRO–STEM CELL ARGUMENTS

By far the most common pro–stem cell argument is that derivation of human embryonic stem cells is morally innocuous because human embryos have no moral status. By analyzing their characteristics, we can see that they are not persons and that it is not wrong to kill them, and we can also see that they do not qualify for any kind of "lower" moral status (for instance, based on sentience). Given that a few other provisos are fulfilled (e.g., concerning permission from the owners of the embryo or gametes), it can then be shown that derivation of stem cells is morally acceptable, perhaps even mandatory. I call this "the standard liberal argument."

One main problem with this argument is that it proves far too much. This is not a new insight, and I do not want to claim any originality for it, but it is worth reiterating because the full consequences of the argument are often suppressed in the public debate. The standard argument proves too much in a number of directions.

First, and perhaps most important, it justifies the (nonpainful) killing and use of any prepersonal human entity from the fertilized egg to the prepersonal infant. Such a killing can be justified by any kind of net benefit to others. In the current context, it can therefore just as easily justify the killing of infants for their stem cells as it can the destruction of embryos for the same purpose. There is no in principle difference between the two killings.[2]

Second, it places no restrictions on the use of biological material from prepersonal human entities that can justify the destruction of these entities, as long as those uses are beneficial. The derivation of a new and effective antiwrinkle cream can therefore be a perfectly acceptable justification for the production and destruction of embryos (or fetuses or infants, although the price would probably be quite high for the infant-derived variety). The attraction of the standard liberal argument seems to be twofold: (1) it solves a whole range of bioethical problems in one fell swoop, from embryo research to euthanasia; and (2) it is attractively economical by locating the wrongness of killing in one and only one consideration. One may, however, wonder whether it is not the result of a misapplication of Occam's razor. If moral values and judgments supervene on natural features of the world, do we then have any reason to believe that the supervenience relationship is of the exceedingly simple kind that personhood theories posit? If I tried to claim that the beauty of a painting was only, or even primarily, due to how the color black had been used (i.e., if I claimed that the only part of the painting important to the supervenience relation between the physical painting and our perception of it was the black parts), I would rightly be taken to task for this ludicrous simplification. But are personhood theories not in the same boat by claiming that there is one and only one feature of any killing that makes it wrong?

A further problem is that the standard liberal argument normally hides some real differences between personhood theorists. They may all agree that embryos are not persons, but at the other end of personal human life, in infancy, they advocate slightly different criteria for the ascription of personhood. One set of criteria will therefore permit the killing of a human individual that, based on another set of criteria, would have been strongly prohibited as the killing of a person. Personhood theorists are also wont to claim that personhood is a threshold concept. This threshold claim can be understood in two different ways: first, as a claim that either you are a person or you are not a person and there are no partial persons; and second, as a claim that during the development of an infant personhood is attained at some point and is then a stable trait. The first of these claims does not entail the second nor vice versa. This is important to realize because it could mean that infants go in and out of personhood a number of times during their development. Let us, for instance, say that personhood is dependent on knowing that you have a life (existence over time) and valuing that life (existence over time).[3] This is clearly a fairly high-level mental trait or ability, and if it develops in a way similar to other high-level traits or abilities, it will not just appear one day and then be present (even as a capability) from then on. Other high-level traits or abilities appear, then disappear, then appear again, and so on until they are finally permanently established.[4]

Another type of pro–stem cell argument tries to bypass the question of moral status by showing that stem cell derivation is justified on direct consequentialist grounds. If the good that can be attained is of a sufficient magnitude, then it can outweigh the killing of a certain number of human entities. It is justified to sacrifice some for the benefits of others (as numerous variations of the trolley example as supposed to show.[5])

This kind of argument may again prove too much—for instance, that it is also justified to sacrifice some adults if the benefits to others are sufficient.

It does, however, also suffer from another problem caused by the embryonic state of present stem cell research. The benefits that are put into the balance to justify the sacrifice are mainly the therapeutic potential promised by stem cell therapy. The public presentation of the benefits of stem cell research has often been characterized by the promise of huge and immediate benefits. As with many other scientific breakthroughs, the public has been promised real benefits within 5–10 years (i.e., in this case, significant stem cell therapies in routine clinical use).[6] Several of the 5–10 years have now elapsed, and the promised therapies are still not anywhere close to routine clinical use.[7] There are similarities to the initial enthusiastic presentation of gene therapy in the late 1980s, and the later problems encountered, and some reason to fear that stem cell therapies will have an equally long trajectory between theoretical possibility and clinical practice. It is likely that many of the current sufferers from some of the conditions for which stem cell therapies have been promised will be long dead before the therapies actually arrive.[8] We are also in a situation where we do not know which, if any, of the possible lines of stem cell research is going to lead to the therapeutic breakthroughs.

This situation can also be modeled by a trolley example. Our trolley is hurtling down the track, and we are approaching a junction where we have to choose between two tracks that both disappear into the distance. On track 1, there are, at what looks like regular intervals, people lying on the track whom we will kill, and there are injured and suffering people sitting at the side of the track whom we will not be able to help. On track 2, the same is the case but there are even more people lying on the track. There is, however, a large sign saying that there is a possibility that we may learn how to stop the trolley somewhere along the track and proceed in a way so that we can both save those lying on the track and help those sitting at the side. Which track should we choose? On track 1, we will be killing and leaving the same number unhelped, but we will have the possibility that at some point this will be made good by our learning to stop the trolley and proceeding in a helpful manner. One track has a certain outcome, one an outcome that may be better or may be worse than the baseline certain outcome. Our assessment of what to do must depend on a number of factors, including (1) our assessment of the probability of learning how to stop and of the time span before we have learned this, (2) our assessment of the probability that other things will happen that will remove some people from the track and help the needy at the side, (3) our temporal discount rate, and (4) our degree of risk aversion.

Now, it may be tempting to claim that risk adversity should play no role in rational decision making,[9] but, given that the probabilities are not strictly quantifiable, denying a separate role to our attitudes to risk will only mean that we will change our (supposedly unbiased) probability estimates to include this attitude component.

In the situation being contemplated here it would be rational to choose track 2, the "uncertain" track, if in our assessment probability 1 was sufficiently high, probability 2 sufficiently low, and we are risk neutral or not very risk averse. There will thus be a fairly large cluster of estimates of these probabilities that make it rational to choose track 2.

Let us, however, complicate matters by imagining that we are faced with not a two-way choice but a three-way choice. There is yet another track, track 3, with the same number of people on the track and at the sides as on track 1, the "certain" track. However, on track 3, a large sign says that there is a possibility that we may learn how to stop the trolley somewhere along the track and proceed in a way so that we can both save those lying on the track and help those sitting at the side. It further states that the probability involved may be less than the probability on track 2, but that the information needed to say whether it is less and how much is unavailable at the time we have to choose.

Given that track 3 is no worse and may be better than the first, "certain" track, we know that we should never choose track 1, but the choice between the two uncertain tracks becomes very complicated and will depend critically on our assessment of the different probabilities involved.

Because I am modeling here the current state of stem cell research, we do, however, have a further option. We can send more than one trolley down the track. Should we send two trolleys down tracks 2 and 3? The choice between sending a trolley down track 3 only and sending trolleys down tracks 2 and 3 will crucially depend on our assessment of whether one of the following scenarios is likely to occur: (1) our progress in finding a way to stop the trolley on track 2 will be so much faster than on track 3 that it outweighs the initial extra loss of life; or (2) on track 2 we will learn things about helping the needy that we will never learn on track 3. Let us translate this back to stem cell research, where track 2 is destructive research and track 3 all kinds of nondestructive research. If embryos have moral status, our pure consequentialist analysis shows that we should pursue only destructive research (on any kind of human beings) if either (1) this line of research will produce valuable therapeutic results much faster than any of the different nondestructive lines of research, or (2) destructive research will be the only way to produce cures or treatments for certain diseases. At present, I submit that we have no strong reasons to believe that any of these two scenarios are likely to be the case and that the pure consequentialist argument therefore at present supports a restriction of stem cell research to nondestructive research.

THE ANTI–STEM CELL ARGUMENTS

The direct arguments against the derivation of human embryonic stem cells, the anti–stem cell argument, do, like the standard liberal argument, focus on the moral status of the embryo.

The standard restrictive argument proceeds from the premise that the life of human individuals is intrinsically valuable at all stages of life. Killing or destroying a human being is, therefore, always *pro tanto* wrong (or, to put it differently, the act always contains at least one wrong-making feature).

The most radical version of the standard restrictive argument claims that all human beings have the same value. On this view, destroying embryos is just as bad as destroying adults, and it seems to rule out any kind of destructive stem cell derivation from embryos (at least if the direct consequentialist argument discussed above does not go through).

This view leads to a number of counterintuitive conclusions in other areas if it is pursued rigorously. It rules out abortion in most, or perhaps all, cases and would, for instance, definitely rule out abortions in cases of rape. It would also put a stop to all kinds of IVF involving the generation of supernumerary embryos and all destructive embryos experiments aimed at improving IVF. It would, finally, also commit us to doing much more to prevent loss of embryos or fetuses during normal reproduction.

Others, including myself, have argued that, although human life is intrinsically valuable at all stages of life, it generally becomes more valuable during the development from fertilized egg to adult human being. This is either argued on the basis of potentiality or on the basis that as the developing human being acquires new characteristics the act of killing it entails more and more wrong-making features (because it deprives it of more and more of what it has, as well as of everything it will get in the future).[10]

It is a commonplace in the "liberal" literature to discount potentiality without any serious consideration, but the arguments used are often exceedingly superficial and build on a conflation of potentiality and mere logical possibility.

On a gradualist analysis of the moral status of the human being through its developmental stages, destroying embryos is always wrong to some degree and cannot be done just for any kind of benefit. One great advantage of a gradualist analysis applied to stem cell research is that it can explain why the destructive use of embryos for stem cell production is less problematic than the destructive use of fetuses, or infants.

But if embryos have value, the question is whether the benefits of stem cell research are of the right kind and large enough to justify the destruction of some valuable embryos. The benefits definitely seem to be of the right kind. Helping or perhaps even curing people who are suffering from horrendous diseases is clearly an ethically worthwhile activity and a good goal to pursue.[11] But what about the magnitude of the benefit; is it also large enough?

One way of answering this question is by finding some other analogous activity where we sacrifice embryos to produce some good. It has been argued by my colleague John Harris that "normal" reproduction is such an activity.[12] For every pregnancy brought to term, on average 3–5 embryos fail to implant or undergo spontaneous abortion some time during pregnancy. Harris argues that, if we accept that loss of embryos by engaging in reproduction (or just not protecting against reproduction), we are forced by reasons of consistency also to allow similar sacrifice of embryos for goals of equal importance. Let us for the sake of argument accept Harris's analysis.[13] How much embryonic stem cell research would it commit us to allowing?

If the trade-off is one life produced for every six embryos used, probably not very much. First, we have to remember that we cannot just count lives saved by stem cell therapies and assume that if we save one life for every six embryos destroyed we will be ethically all right. We can save the same life many times, and it is questionable whether every new saving counts equally. If I am hand-ventilating a patient, do I save his life anew with every compression of the ventilation bag?[14] We also have to take account of the fact that many stem cell therapies are not lifesaving but life extending or quality-of-life enhancing. We therefore need a metric that can take account of all these considerations. One such metric is the QALY (quality adjusted life year). QALYs have many problems, but none of them should be of relevance to the following analysis.

Assuming a low-average life span of 72 years for a newborn, the "normal" trade-off works out as 12 life years per embryo, or, if we assume that some of these are of lower quality than others so that the net QALY gain per newborn is only 60 QALYs, we still have a trade-off of 10 QALYs per embryo.

A recent paper from one of the well-known groups in the field reports the generation of one definite and two possible embryonic stem cell lines from 50 blastocysts (and 101 fertilized eggs).[15] If we assume that all the cell lines are true embryonic stem cells, this is equivalent to a 1 in 16.6 ratio between embryo use and cell line production.

If all three cell lines turned out to be therapeutically useful, it would thus seem that they would each have to produce health improvements equivalent to at least 166 QALYs to be equivalent to normal human reproduction.[16] It is, however, unlikely that all the cell lines generated will ever have any therapeutic usefulness. Many will only be used for basic research, and others will be discarded because they are less easy to work with than other similar lines. It might also be the case that the use of adult stem cell lines derived in nondestructive ways will become the preferred option in a number of therapeutic areas, or we may be able to create stem cells without the involvement of embryos, both developments leaving the derived embryonic lines unproductive. We are still at such an early stage of development that we have no way of estimating how many of the stem cell lines that are derived will eventually be used for therapeutic purposes. But, all in all, the considerations here indicate that on this type of consistency argument the stem cell lines that are actually used would have to produce major therapeutic benefits.

As discussed in the first section of this [chapter], the standard liberal argument has a number of rather embarrassing consequences if it is really taken seriously. The same is true of the standard restrictive argument discussed in this section, even in its gradualist version. If embryos have significant moral status, it would commit us to a much stricter abortion policy than many find acceptable, although it would be possible to allow abortions in certain circumstances, and it would also commit us to rethinking our research priorities with regard to embryo and fetal loss during normal pregnancy.

THE POLICY STALEMATE

As I have discussed, each side in the debate on stem cells has a number of rather nasty skeletons in their ethical cupboard that are usually hidden from public view. The full implications of the views are not explored in public debate, and if they are pointed out by opponents they are often dismissed as irrelevant to the current debate. This unfortunately leads to a dishonest public debate. The participants are perhaps on analysis minimally truthful (i.e., they do not lie), but they nevertheless argue in a way that is (intentionally?) misleading.

These features of the debate contribute to the policy stalemate that can be discerned in every country where public policy has been formulated on human embryonic stem cells. Neither side of the debate has a policy that is consistent with the full implications of its argument,[17] and neither side seems very interested in pressing for such a fully consistent policy!

This situation should perhaps in itself make us wary of trying to decide public policy based purely on consistency arguments (because they seem to be very selectively employed), but let us nevertheless as good philosophers try to see how far consistency will get us when applied to the regulation of stem cell research.

When consistency is invoked concerning public policy, the claim is that certain features of current regulation in a specific area logically (or perhaps slightly weaker, morally) commit regulators to regulate new areas in a specific way for reasons of consistency or parity of reasoning.[18] This type of argument occurs in two subtypes. The first subtype is:

1. Authority X promulgates regulation R.
2. X puts forward an incomplete argument A supporting R.
3. A can be made complete by the addition of premises $P_1 - P_n$.
4. The complete argument A will (perhaps with the addition of some further premises that X accepts) also support regulation C.
5. X is therefore, by parity of reasoning, committed to accept regulation C.

The second subtype is:

1. Authority X promulgates regulation R.
2. The justification for R can be reconstructed as argument A*.
3. The argument A* will (perhaps with the addition of some further premises that X accepts) also support regulation C.
4. X is therefore, by parity of reasoning, committed to accept regulation C.

These "official policy" based arguments may fail if our reconstruction and completion of the underlying arguments are wrong, but a further possible mode of failure occurs if the regulation R has been promulgated in a situation where R is the result of a compromise, either because X has openly engaged in a compromise with some other group or because X has already taken the views of some other group into account when drafting the regulation (for instance, to ensure a smooth passage through the political process). If a regulation is based on a political compromise, it may well be the case that no one wants to defend or justify the position reached in the regulation, except as a legitimate result of a legitimate political process.[19]

An example could be the abortion legislation in countries allowing abortion on demand until a certain gestational age. It is unlikely that many people would actually claim to have an argument that can justify the exact limit legislated in a specific country, but many more might be willing to accept it as a legitimate political compromise between those wanting a higher limit and those wanting a lower limit. Such laws do, therefore, not necessarily imply that something radical is thought to happen to the *moral* status of the fetus or the *moral* rights of the mother when the gestational age in question is reached.

A consistency argument based on official policy also needs to take into account that there are limits to the preciseness with which one can draft regulations and that ethically extraneous factors play a role. As Beauchamp and Childress write:

> Public policy is often formulated in contexts that are marked by profound social disagreements, uncertainties, and different interpretations of history. No body of abstract moral principles and rules can determine policy in such circumstances, because it cannot contain enough specific information or provide direct and discerning guidance. The specification and implementation of moral principles and rules must take account of problems of feasibility, efficiency, cultural pluralism, political procedures, uncertainty about risk, noncompliance by patients, and the like. (pp. 8–9)[20]

In evaluating official policy it is also very important not to forget that the ideal policy has to be both locally and globally coherent and consistent and that maximal coherence in a body of regulation may be achieved by allowing some logical inconsistencies. In many legal systems it is, for instance, accepted that a major change in legal status occurs at birth. This principle permeates many areas of law, also outside of the laws strictly dealing with reproductive matters, and it is at least arguable that global consistency is better served by upholding it in all cases, even if there are a few cases where it might seem to be logically inconsistent.

In the context of stem cell research, one particular consistency argument is often put forward claiming that, because destructive research using embryos for certain purposes is already allowed in a given jurisdiction, consistency requires that we allow the destructive use of embryos for stem cell research.[21]

Many countries that presently allow experiments on embryos restrict these experiments to those related to improving reproductive technologies and increasing our understanding of the biology of human reproduction. Many types of stem cell research fall outside this restriction and are therefore not permitted even though the jurisdictions in question allow other forms of research on embryos.

Can a restriction of embryo research to reproductive matters be justified?

The decision to allow embryo research for a restricted range of research questions could possibly be reconstructed as an attempt to achieve consistency in a situation where embryo research is believed to be (somewhat) ethically problematic.[22] Most legal regulation of embryo research occurs within the broader context of regulation of assisted reproductive technologies, and can only be improved, if embryo research is permitted. Any legislation that allows the use of assisted reproductive technologies and prohibits embryo research could therefore be charged with a form of performative inconsistency, by prohibiting a necessary step in the development and improvement of a permitted technique.[23]

Would it then be inconsistent not to allow research using embryos with no connection to reproduction?

Michael Walzer has famously argued that society contains several separate "spheres of justice" and that the application of principles of justice from one sphere to another is not necessarily warranted.[24] Maybe there are also separate "spheres of consistency" when we discuss the consistency of public regulations. This is not as strange an idea as it might perhaps initially seem.

Walzer's argument for different spheres of justice is that different social goods are different and that, unless we understand and take account of this difference in our analysis of justice, we will go wrong. He writes:

> No account of the meaning of a social good, or of the sphere within which it legitimately operates, will be uncontroversial. Nor is there any neat procedure for generating or testing different accounts. At best, the argument will be rough, reflecting the diverse and con-

> flict-ridden character of the social life that we seek simultaneously to understand and to regulate—but not to regulate until we understand. I shall set aside, then, all claims made on behalf of any single distributive criterion, for no such criterion can possibly match the diversity of social goods.[25]

In the same way as different social goods are different, the different social practices aimed at producing the social goods are different. The practices involved in reproduction and in securing the goods secured by reproduction are different from the practices involved in research. There is thus no prima facie reason to believe that arguments and conclusions valid in one of these areas can be transferred without modification to the other area. Let me try to illustrate this.

In the area of reproductive ethics the idea that there is a strong right to reproductive freedom or reproductive liberty has gained currency in recent years. If a given permissive legal regulation has been passed because of an appreciation of this right (for instance, concerning the creation, use, and destruction of embryos in the sphere of ARTs), then it is not immediately obvious that consistency requires the same kind of permissive regulation outside of the sphere of reproduction. A right to reproductive liberty cannot, for instance, in itself support a permission to use embryos for nonreproduction-related purposes, such as stem cell research! And, given that the right to reproductive liberty is most often justified by arguments pointing to the central position of reproductive projects in the lives of (most) people, it also follows that the consideration justifying a right to reproductive liberty cannot directly support the production of embryos for nonreproduction-related purposes (whether they be research or other purposes).

In looking at consistency arguments in the public policy sphere, we might also want to question whether "as morally acceptable"[26] is a transitive relation in the context. If p is as morally acceptable as q, and q is as morally acceptable as r, have we therefore committed ourselves to the judgment that p is as morally acceptable as r?

Moral acceptability comes in at least two different forms, and the difference between them can be brought out by considering Tranøy's analysis of moral disagreement and consensus. If A and B discuss the morality of p, there are, according to Tranøy, three possible outcomes:

1. Persons A and B can agree that p is *unacceptable,* i.e., both mean *positively* that p should not be accepted,
2. A and B can agree that p is *acceptable,* i.e., both mean *positively* that p should be accepted,
3. A and B can agree that p is acceptable, but A or B might *abstain* from actively taking a stand on p.[27]

Tranøy calls the last option "open consensus."[28]

Let us call the kind of moral acceptability where both decision makers positively affirm the moral acceptability of p "moral acceptability" and the kind of moral acceptability where one of the decision makers abstains from taking a stand on p "moral acceptability*." Let us assume that moral acceptability is transitive; does this entail that moral acceptability* is also transitive? Clearly not. The judgment of moral acceptability* may occur for a number of different reasons. The case p may for B be in a gray area between the clearly acceptable or the clearly unacceptable, or B may defer to A's judgment because the question at issue is much more important for A. Moral acceptability* is thus only transitive in those cases where B's reason for abstaining from actively taking a stand on p is shared by q and r, and, because these reasons may be nonmoral, transitivity of moral acceptability* only occurs if we can claim parity of reasoning for both the moral features and the nonmoral features of p, q, and r.

It should be relatively obvious that the kind of moral acceptability that lies behind public regulation of controversial areas in bioethics is most often not moral acceptability but moral acceptability*. This means that purely ethical consistency arguments are often misapplied, because they assume a transitivity of moral acceptability that often is not there.

CONCLUSION

In this [chapter], I established that the standard liberal arguments for allowing stem cell research are problematic, as are the standard restrictive arguments. For this reason, they cannot be used to guide public policy.

I have further shown that any line of argument trying to circumvent a reliance on the standard arguments by instead relying on consistency to guide public policy formation in this area is also problematic because (1) there are general problems with consistency arguments as they are currently used, (2) there are specific problems affecting consistency arguments in public policy, and (3) there are even more specific problems in seeking consistency between the sphere of reproductive issues (where personal reproductive liberty plays a major role) and the sphere of research and therapy.

These conclusions are all negative but should not, on reflection, be terribly controversial.[29] The positive, but controversial, conclusion is that at least some of these problems can be resolved by adopting a gradualist position on the moral status of the embryo. If we adopt a gradualist position, we will find that we need good reasons to perform any kind of experiments on embryos but that such experiments are not ruled out *tout court*. This means that it is up to the scientists to present convincing arguments that destructive embryonic stem cell research is necessary to produce cures and treatments for important human diseases.

NOTES

1. For a review of the debate, see S. Holms, "Going to the Roots of the Stem Cell Controversy," *Bioethics* 16 (2002): 493–507.

2. We might try to say that, if we want to kill the prepersonal infant for its stem cells, we need the permission of its parents and that the parents are then obligated to take the best interests of the infant into account (and presumably not allow the killing). This may be correct in law, because the law considers every born human being as a person, but it is not morally justifiable on the standard argument. If the infant is not a person, the parents are not proxy decision makers but simply decision makers for a piece of property in which they have invested time and resources.

3. This is close to the Harris and Tooley positions.

4. V. Lee and P. D. Gupta, eds., *Children's Cognitive and Language Development* (Oxford: Blackwell, 1995); J. McShane, *Cognitive Development: An Information Processing Approach* (Oxford: Blackwell, 1991); J. Piaget and B. Inhelder, *The Growth of Logical Thinking from Childhood to Adolescence* (New York: Basic Books, 1958).

5. F. M. Kamm, "The Doctrine of Triple Effect and Why a Rational Agent Need Not Intend the Means to His End," *Proceedings of the Aristotelian Society* S74 (2000): 21–39; J. Harris, "The Moral Difference between

Throwing a Trolley at a Person and Throwing a Person at a Trolley," *Proceedings of the Aristotelian Society* S74 (2000): 40–57.

6. Anonymous, "Taking Stock of Spin Science," *Nature Biotechnology* 16 (1998): 1291.

7. Given the time needed for basic research, clinical research, and regulatory approval, it is unlikely that any therapy using biological materials and based on a truly novel therapeutic approach could move from initial discovery to clinical use in five to ten years. See also R. Lovell-Badge, "The Future of Stem Cell Research," *Nature* 14 (2001): 88–91.

8. B. Albert, "Presentations to the All-Party Disablement Group," unpublished manuscript, July 25, 2000.

9. This is, for instance, a central point in the argument between John Rawls and his consequentialist critics about his argument that a rational decision maker behind the veil of ignorance would choose the maximum decision to the distribution of basic goods.

10. S. Holm, "The Moral Status of the Prepersonal Human Being: The Argument from Potential Reconsidered," in *Conceiving the Embryo*, ed. D. Evans (Dordrecht: Kluwer, 1996), pp. 193–220.

11. Whether the same can be said about basic biological research using human embryonic stem cells is another matter.

12. J. Harris, "The Ethical Use of Human Embryonic Stem Cells in Research and Therapy," in *A Companion to Bioethics*, ed. J. Burley and J. Harris (Oxford: Blackwell, 2002), pp. 158–74.

13. Most gradualists would probably argue that embryos have more value than one-sixth of an adult life and that the reason we "allow" loss of embryos in reproduction is not primarily connected to the value of the embryos.

14. This issue is quite complex, and intuitions may vary from case to case. Here is one more case supporting the line I take on repeated savings. Let us imagine that by killing one person and removing his organs we can get five organs that are lifesaving for others. This might, on a utilitarian analysis, justify the killing as shown by John Harris in his famous survival lottery paper. But would it also justify killing the one person in a situation where we knew that a person would need the five organs sequentially over a period of twenty-five years—that is, where we had five lifesaving episodes, but all of these were lifesaving for the same person? In my book, this would be to kill one person to save the life of one other person—not killing one to save five.

15. S. E. Lanzednorf et al., "Use of Human Gametes Obtained from Anonymous Donors for the Production of Human Embryonic Stem Cell Lines," *Fertility and Sterility* 76, no. 1 (2001): 132–37.

16. This, by the way, indicates that it is quite unlikely that this consistency argument could ever show individually targeted stem cell production (for instance, by nuclear replacement) to be ethically acceptable.

Unless longevity increases enormously, it will simply be close to impossible to generate the requisite number of QALYs or unadjusted life years. This will be the case even if you could use the same cell line for a number of different therapies in the same person. It will still be very difficult to cram 166 QALYs into one life.

17. It is important to note that this is true both in restrictive and liberal justifications. The restrictive regulations of the United States are not nearly restrictive enough and the liberal regulations of the United Kingdom not nearly liberal enough to match the philosophical argument.

18. The following section is a slightly amended sections from my paper "Parity of Reasoning Argument in Bioethics—Some Methodological Considerations," in *Scratching the Surface of Bioethics*, ed. M. Häyry and T. Takala (Amsterdam: Rodopi, 2003), pp. 45–56.

19. It is usually accepted that a political decision can be legitimate even in cases where it can be argued to not be optimal.

20. T. L. Beauchamp and F. J. Childress, *Principles of Biomedical Ethics*, 5th ed. (New York: Oxford University Press, 2001).

21. Harris, "The Ethical Use of Human Embryonic Stem Cells in Research and Therapy." The possible policy responses to stem cell research also raise a number of other consistency issues that will not be discussed in detail here. These include questions of consistency with regard to (1) legal permission of certain kinds of stem cell research and prohibition of public funding of the same kinds of research, and (2) legal prohibition of derivation of stem cell lines from embryos and permission to import and use cell lines derived elsewhere.

22. Many public policies seem to indicate that people are "closet gradualists" with regard to the moral status of embryos and fetuses.

23. It is unclear whether any formal inconsistency would be implied. It is still an open question whether subsequent use of research data obtained in unethical experiment (e.g., the Nazi concentration camp experiments) is in itself ethically problematic.

24. M. Walzer, *Spheres of Justice: A Defense of Pluralism and Equality* (Oxford: Blackwell, 1983).

25. Ibid., p. 21.

26. Or the many similar locations like "as ethically justified."

27. K. E. Tranøy, *The Open Mind: Morals and Ethics in a New Millennium* (Oslo: Universitetsforlaget, 1998), p. 155. I owe the translation from Norwegian to Jan Helge Solbakk.

28. Open consensus can also occur around the judgment from p is unacceptable if either A or B abstains from actively taking a stand on p.

29. Even the most insistent liberal and conservative philosophers usually shy away when asked whether they want to implement the full implications of their position as social policy.

16

Stem Cell Research

The Failure of Bioethics

Don Marquis

In recent years the issue of human embryonic stem cell (hereafter HESC) research has engendered fierce debate. Some object to HESC research because they say it involves taking a human life. Others argue that its prospective benefits are so huge that not to pursue it would be immoral. It is unreasonable to think that such a controversy will be resolved by journalists or politicians or, for that matter, by patients who hope for a cure from some dreadful disease.

However, it does seem reasonable that practitioners of academic bioethics should be able to help us clarify this issue. Presumably academics have the proper interests and education to think clearly about bioethics controversies. Academic essays provide an opportunity for careful examination of relevant arguments. Since leading bioethics and medical journals have devoted whole recent issues or parts of recent issues to this controversy,[1] examining this literature should shed considerable light on the HESC controversy.

Such an expectation, in this reader's experience, will be disappointed. It is amazing how much of the HESC bioethics literature

Reprinted from *Free Inquiry* (Winter 2002–2003): 40–44.

confines itself to describing the science underlying the dispute, or to giving historical accounts of the committees that have reported on this issue, or to surveying the dispute in general terms rather than closely analyzing the arguments that bear on the central ethical issues.[2] The arguments that are offered are typically presented in a cursory way. Indeed, often they are more suggested than presented. Arguments that deserve critical scrutiny are quickly set out as if any rational reader would regard them as obviously sound.

No doubt there are many explanations for this. One, however, may have particular relevance to the HESC issue. Obtaining HESCs requires the destruction (or *disaggregation*, to use the sanitized term) of human embryos.[3] Whether such embryo destruction is morally permissible is (or should be) at the heart of the debate over the morality of HESC research. This has led many to expect that to a great extent, the stem cell controversy will mirror the abortion controversy. If one believes there are good arguments for the moral permissibility of fetal destruction, then many of the same arguments can apply to embryo destruction. When these considerations are combined with the fact that most partisans on both sides of the abortion dispute seem to consider their positions so obviously true that only cursory argument in defense of them is needed, we may have a plausible explanation for the superficial nature of the HESC discussion.

The purpose of this [chapter] is to provide evidence for the claims made in the above two paragraphs. Consider first the major arguments for this view that HESC research should be banned, or at any rate not funded. Richard Doerflinger has defended this view on the grounds that it is incompatible with a Catholic viewpoint.[4] Gilbert Meilaender has defended the view on more general religious grounds.[5] There are obvious problems with such defenses. Religious views are essentially matters of faith, and it is widely thought that there are not objective grounds for preferring one variety of religious faith to others. The fact that there are so many varieties of Christianity and of Islam and, I suppose, of many other religions of which I know little, is evidence for this. Accordingly, in the absence of a great deal of convincing argument that has not yet seen the light of day, particular religious considerations cannot establish the wrongness of HESC research. Such religious views count as no more than comforting opinions, like preferences in furniture or food, and we are left with no reason whatsoever to accept them as binding on the rest of us, or for that matter, even binding on their proponents.

No doubt, the weakness of the religious arguments has suggested to some proponents of HESC research that defending their position requires only a wave at some nonreligious arguments. This may count, therefore, as another reason arguments concerning HESC research seem less than compelling.

The letter that invited this [chapter] stated that many readers of [*Free Inquiry*] hold a view that, if it were true, might furnish a good argument in favor of HESC research. According to the editor, many *Free Inquiry* readers "support untrammeled freedom in scientific research." Yet to accept this slogan entails endorsement of the Tuskegee studies of the natural course of syphilis, or the notorious Willowbrook experiments in which retarded children were deliberately given hepatitis in order to study how that disease progressed, or the Nazi hypothermia experiments on Jews. However, there is a consensus among decent persons of all political and religious persuasions that these studies were profoundly immoral. Thus such an "untrammeled freedom" argument is plainly unsound.

Another argument for HESC research is based on the hope that it might provide cures, or at least reasonably successful treatments, for Alzheimer's disease, spinal cord injuries, muscular dystrophy, diabetes, and other diseases. This prospective benefits argument, although often offered in popular or political discussion, is as hopeless as the untrammeled freedom argument. The Tuskegee, Willowbrook, and Nazi studies were wrong, not because they were bad or useless science, but because the human subjects in them were treated inhumanely. Basic interests of those subjects were sacrificed without their consent for the ends of the research. There is now a strong consensus, both in society and in academic bioethics, that this is wrong even when the research clearly will promote the common good. In short, conformity with a respect for human subjects (RHS) principle is a necessary condition of morally permissible research, whatever its benefits.[6]

Just how should this RHS principle be formulated and qualified? Here is an issue that can provide a clear framework for analysis of the HESC dispute. Critics of HESC research claim in essence that it violates the principle of respect for human subjects. One can make at least a presumptive defense of the truth of this claim in the absence of any appeal to religion. Destroyed human embryos clearly belong to our species. Therefore, they were *human* in one clear sense. Furthermore, the embryos in such research are clearly *subjects*, that is, the subjects of research. Thus, as the more thoughtful proponents of HESC research at least tacitly realize, arguments in favor of such research must appeal, not to an *unrestricted* or *unqualified*, RHS principle, but to a *restricted* or *qualified*, version. The case for HESC research requires the defense of a restricted version of the RHS principle that, on the one hand, does not include embryos and, on the other hand, includes those human subjects whose need for protection is uncontroversial.

How might such a case be made? Some authors have suggested that embryo destruction is permissible because embryos are not moral agents.[7] This in turn suggests a respect principle restricted to human subjects who are also moral agents. Plainly this will not do, since it is agreed by all decent people that scientific research that involves the destruction of two-year-olds is immoral, even though two-year-olds are not moral agents. It is worth noting, in this connection, that those retarded children who were deliberately infected with hepatitis in the Willowbrook experiments were not moral agents either.

Other authors have suggested that HESC research is morally permissible because embryos are not persons.[8] This is tantamount to claiming that the respect principle should be restricted to persons. A relevant meaning of *person* uses the term to refer to a cluster of psychological characteristics that human beings typically possess and animals do not.[9] But this restriction on the RHS principle is rejected by almost everyone. Infants clearly are not persons in this sense, and yet we believe that research that causes the destruction of infants is immoral.

One proponent of HESC research has offered a response to this objection. Carson Strong has defended the view that, although infants are not persons in virtue of any intrinsic property, good reasons exist for *conferring* the right to life on them because of the consequences for persons of doing so. Treating infants with love and concern will have good consequences for the persons they grow up to be. It will also have good consequences for other persons who may be affected adversely by those human beings who were treated badly when they were infants. Furthermore, treating some infants as expendable might lead us to lack concern for infants we intended to keep.[10] According to Strong these considerations, combined with the fact that infants "are *viable, sentient,* have the *potential to become self-conscious,* have been *born,* and are *similar in appearance* to the paradigm of human persons" are "significant enough to warrant conferring upon infants serious moral standing including a right to life."[11]

Strong claims that, on a moral theory in which only persons have full moral status in and of themselves, *our* conferral of full moral status on infants is mandated. This is hard to believe. All of the reasons Strong cites for treating infants with love and concern apply as well to fetuses. Yet many believe that killing a fetus a woman does not intend to keep is perfectly compatible with love and concern for a fetus she does intend to keep. Strong might protest that this objection does not take account of the fact that infants are viable, sentient, have been born, and are like paradigm humans in appearance, whereas fetuses are not. The trouble with Strong's view is that he argues that none of the more plausible of these factors is sufficient to underwrite full moral status. Given that, why should we assume that collectively they are of much help? Strong offers no argument here, only assertion.[12] Thus, we are left with good reason thinking that restricting the RHS principle to persons is far too narrow.

Some have suggested that the destruction of embryos is justified because embryos are not capable of sentience, that is, they lack the capacity for thoughts, feelings, or experiences.[13] For this argument to go through, the respect principle must be restricted to individuals with a capacity for consciousness. Embryos lack present capacity for consciousness. So do temporarily unconscious adults. We all agree that research on temporarily unconscious adults that causes their death is immoral. Hence, the present capacity for consciousness restriction on the RHS principle is untenable.

This problem with temporarily unconscious adults can be fixed by adopting a RHS principle that excludes only those in whom the absence of consciousness is *irreversible*. Because it is reasonable to think that a subject whose absence of consciousness is irreversible cannot be harmed in any morally significant way by his or her destruction, there is a good deal to be said for this change. However, an embryo's lack of consciousness is not irreversible. In the proper environment embryos may develop mental capacities like our own. On this interpretation, a respect principle restricted to those with the capacity for consciousness (or a first cousin) might be defensible, but it does not exclude embryos.

One might attempt to deal with both of these problems by claiming that the RHS principle should be restricted to those who have in the *past* exhibited the capacity for consciousness. Critics of HESC research should respond that, instead, the RHS principle should be restricted to those who could in the *future* exhibit the capacity for consciousness. How should we resolve this disagreement? Society endorses medicine's concern with *prognosis*. In view of this, critics of HESC research seem to have the more defensible view.

An argument often given in defense of HESC research is that an embryo "is not clearly even an *individual*,"[14] that at the embryo stage, "it is doubtful one can speak of individuality."[15] Carol Tauer has reported that a consensus of the National Institutes of Health Human Embryo Research Panel favoring HESC research found that the embryo is not "a distinct individual," that it lacks "developmental individuation" and thus lacks full moral status.[16] Ronald Green has made essentially the same argument.[17] Restricting the RHS principle to individuals certainly seems reasonable. But why should we suppose that embryos are not individuals?

Here is a reason for supposing that they are. One can count them. One can unambiguously refer to one embryo, and then to a second, and then to a third. So it seems that embryos are indeed individuals and that restricting the RHS principle to individuals does not succeed in excluding embryos from its scope. Supporters of HESC research will protest that embryos are not individuals because of the possibility of twinning. But why should one suppose that this is a reason that embryos are not individuals? One amoeba can split into two. This should not tempt us to claim that it was not the case that there was one individual amoeba before the splitting. Indeed, how could one understand what *splitting into two* is unless the amoeba were an individual before the split? But unless one understands what splitting into two is, one cannot understand at all what it is for an embryo to twin. Therefore, the individuality restriction, although defensible, seems not to be helpful to the proponent of embryonic stem cell research.

Supporters of HESC research have also made the argument that:

> If an embryo is maintained outside a woman's body and those who provided the gametes for it have not decided to permit its development in a womb, it is not effectively a state in the early development of a person. Put differently, an extracorporeal

> embryo—whether used in research, discarded, or kept frozen—is simply not a precursor to any ongoing personal narrative.[18]

For this claim to do the work its proponents want it to do, the RHS principle must exclude cases where we fail to provide a human subject with an environment in which it can develop into a person. Thus such a restriction allows research on infants in a neonatal intensive care unit that results in their destruction or, for that matter, on any abandoned infant whatsoever. Accordingly, such a restricted version of the principle is too narrow.

Jeffrey Spike has argued that HESC research is morally permissible because it is not the case that embryos can develop independently.[19] This will hardly do, for the restriction on the RHS principle that is needed to make Spike's argument go through would allow unethical research on the disabled.

Spike has also defended HESC research on the grounds that "biologically if every one of those embryos was put into a woman, perhaps 10 or 20 percent would survive to birth." This won't do either. Research on patients with cancer in which the survival rate is only 10 to 20 percent is not exempt from the RHS principle.

This concludes my survey of the argument that human embryo destruction is morally permissible, or, alternatively, that the principle of respect for human research subjects should be restricted so that embryos are excluded. What has been surveyed is not a pretty picture. In spite of the vast amounts of ink that academic bioethicists have spilled over the HESC issue, and in spite of the apparent sentiment among so many of them that HESC research should proceed, the crucial arguments offered to support this view are both sketchy and subject to all sorts of fairly obvious difficulties. Perhaps there are no arguments that can justify proceeding with HESC research. But that would oblige academic bioethicists to acknowledge the fact, not to pretend that this unpleasant little difficulty does not exist.

It is possible to imagine objections to the proceeding analysis. For example, one might say that embryos cannot possible be included in the RHS principle because that principle concerns the basic interests of human subjects—and embryos, lacking the capacity for consciousness, cannot have interests at all.[20] The argument for the latter claim is that whatever lacks the capacity for consciousness cannot take an interest in anything and what cannot take an interest in anything cannot have interests.[21] This objection is subject to a number of difficulties. The notion of what is in one's interest should not be too tightly associated with what one takes an interest in. Not everything that people take an interest in is in their (best) interest and not everything that is in people's best interest is what they take an interest in. Smokers and drug addicts are obvious examples. Furthermore, the objection cannot account for acting in the best interest of a temporarily unconscious adult or an infant.

Someone also might object to the preceding analysis on the grounds that respect for an individual is compatible with destroying her.[22] It would follow that my interpretation of the respect for human subjects principle is too strong. However, whether or not such a notion of respect is possible is beside the point. The respect for human subjects principle used in standard medical research ethics does not permit harming human subjects without their consent if that harm can be anticipated.

One might also object to my analysis on the grounds that it holds proponents of HESC research to standards that are overly strict. Meyer and Nelson say:

> Our goal, however, is not to provide a knock-down argument about the moral status of the embryo, but to show how one systematic, reasonable view on moral status in general can be used to defend the moral property of destroying embryos that truly deserve respect.[23]

Meyer and Nelson refer to Mary Anne Warren's "developmental view" of human moral status.[24] Other defenders of stem cell research also have appealed to her view.[25] Warren is best known for her view that abortion is morally permissible because fetuses lack full moral status.[26] Warren's developmental view is indeed widely regarded as reasonable—precisely to those who consider abortion morally permissible on the grounds that fetuses lack full moral status because they are only developing human beings and not yet persons. If one holds this, then, of course, one will hold that a human being at the very earliest stages of development lacks moral standing. The triviality of this move boggles the mind. It has all the force of John Paul II's defense of his views of the morality of abortion and euthanasia in terms of—well—his own moral theology.[27] What is required for the success of the Meyer-Nelson point is a criterion for the reasonableness of a view that can be accepted by any rational person.

An adequate analysis of Warren's developmental view is far beyond the scope of this [chapter]. In my view, a theory such as hers that bases full moral standing on moral agency cannot account even for the moral standing of adolescents, much less younger human beings, without some backing and filling and other moves that are utterly arbitrary. If proponents of HESC research are willing to tolerate arbitrariness, then they should not object to what they perceive as the arbitrariness of religious perspectives.

Another objection to the proceeding analysis might be that the RHS principle includes embryos only if the embryo's status as a potential human being gives it the same moral status as an actual human being.[28] It is clearly not wrong to destroy isolated human cells. The only difference between a zygote and some arbitrary skin cell is that the zygote is a potential human being. Any why should we suppose that being a merely potential human being is sufficient to underwrite full moral status?

The critic of HESC research can reply by holding that embryos are *actual* human beings. They are *very, very young* human beings. Not all human beings look like middle-aged professors. It seems doubtful that the ordinary notion of actual human being is sufficiently precise to underwrite either this objection or the response to it.

Another objection to my analysis might concern its strategy. I considered candidate restrictions on a RHS principle individually, and argued that each is indefensible or does not exclude embryos. One might argue that each of those qualities (except for individuality) should be considered to be "candidate sufficient conditions" for full moral standing. One might go on to argue that since embryos meet none of the candidate sufficient conditions for full moral standing, then there is not reason to think that they have moral standing. And if there is no reason to think that they have moral standing, then HESC research is morally permissible.

The trouble with this objection is that in order for it to be successful, one would have to show that all of the candidate sufficient conditions for full moral standing had been considered. In the first place, this has not been done. In the second place, it is easy to think of candidates for full moral standing that proponents of HESC research have not considered.

Another objection to the preceding analysis is that I have presupposed that morality must be objective in a way that it cannot possibly be, and that therefore I have held the defenders of HESC research to impossibly high standards. Such a critic might argue that morality is a social construct, that it is not based on some natural property or other of individuals.[29] This objection opens a very large can of worms. One problem with a somewhat subjectivist conception of morality is showing that it does not permit too much; that in the course of showing that research on human embryos is morally permissible, it does not also show that practices that seem clearly immoral are also morally permissible. That is a difficult, and to my knowledge, not successfully attempted task.

This [chapter] has endorsed no positive thesis. It has been concerned almost exclusively with criticizing the arguments of others. Can something positive and nonreligious be said in favor of banning human embryonic stem cell research? Here's a suggestion for an argument: Failing to respect the basic interests of ordinary human beings for the purpose of scientific research is wrong. Age discrimination is morally wrong. When we were very, very young, we were mere embryos. Therefore, destruction of human embryos for the purposes of scientific research is wrong.

Carson Strong would object to this argument on the grounds that embryos do not *become* self-conscious beings, they only *produce* beings that are capable of self-consciousness. His reason for this claim is that the embryo that was your precursor was also the precursor of the placenta that supported you.[30] Such an argument appears to require the assumption that an entity cannot shed some of its parts and remain self-identical. However, this seems false: consider amputees.

I would not pretend that the preceding suggestion for an argument is anything more than that. Furthermore, even I find the claim that embryos deserve the same moral respect as adult human beings counterintuitive. Nevertheless, because I think that on issues like this, the moral intuitions of another are not authoritative, it would be outrageous for me to believe that my own intuitions are more authoritative. I'm inclined to think that some argument or other concerned with individuality might be successful in showing that HESC research is permissible. However, I used to think that arguments concerned with the individuality issue were much better than they actually are, so I have no confidence at all in this conjecture. Furthermore, the dismal failure of the arguments of so many who think that HESC research is morally permissible suggests that HESC research is not morally permissible. Of course, this conclusion could be shown to be false by only one good argument from HESC proponents.

NOTES

1. Here are some examples: *JAMA* 284, no. 24 (December 2000); *Hasting Center Report* 31, no. 1 (January/February 2001); *Kennedy Institute of Ethics Journal* 9, no. 2 (1999); *American Journal of Bioethics* 2, no. 1 (2002); and *Journal of Medicine and Philosophy* 22, no. 5 (October 1997).

2. A recent issue of *American Journal of Bioethics* (2, no. 2) devoted to the stem cell issue is especially noteworthy in this respect.

3. This wonderful term is found in Robert P. Lanza et al., "The Ethical Validity of Using Nuclear Transfer in Human Transplantation," *JAMA* 284, no. 24 (December 27, 2000): 3175–79.

4. Richard Doerflinger, "The Ethics of Funding Embryonic Stem Cell Research: A Catholic Viewpoint," *Kennedy Institute of Ethics Journal* 9, no. 2 (1999): 137–50. Doerflinger does offer some nonreligious considerations, but these are so cursorily presented that it is unclear exactly what they are. See p. 139.

5. Gilbert Meilaender, "The Point of a Ban, Or, How to Think about Stem Cell Research," *Hastings Center Report* 31, no. 1 (January–February 2001): 9–15.

6. I assume that some authors who appear to reject this principle, such as Glenn McGee and Arthur Caplan, "The Ethics and Politics of Small Sacrifices in Stem Cell Research," *Kennedy Institute of Ethics Journal* 9, no. 2 (1999): 151–58, and Ronald M. Green, "Determining Moral Status," *American Journal of Bioethics* 2, no. 1 (2002): 20–30, are instead committed to qualification of the principle.

7. This is suggested in Michael J. Meyer and Lawrence J. Nelson, "Respecting What We Destroy: Reflection on Human Embryo Research," *Hastings Center Report* 31, no. 1 (January–February 2001): 18.

8. Ibid. Carson Strong seems to hold this view. See his "The Moral Status of Preembryos, Embryos, Fetuses, and Infants," *Journal of Medicine and Philosophy* 22, no. 5 (October 1997): 457–78.

9. Dan Brock uses the term this way in *Life and Death* (Cambridge: Cambridge University Press, 1993), p. 372. Others do also.

10. These arguments can be found, as Strong notes, in S. I. Benn, "Abortion, Infanticide and Respect for Persons," and Joel Feinberg, "Potentiality, Development and Rights." Both are in *The Problem of Abortion*, 2d ed., ed. Joel Feinberg (Belmont, Calif: Wadsworth, 1984). See Strong, "The Moral Status of Preembryos, Embryos, Fetuses, and Infants."

11. Strong, "The Moral Status of Preembryos, Embryos, Fetuses, and Infants," p. 468.

12. Ibid.

13. This argument can be found both in Lanza et al., "The Ethical Validity of Using Nuclear Transfer in Human Transplantation," p. 3177, and Meyer and Nelson, "Respecting What We Destroy: Reflection on Human Embryo Research," p. 18.

14. Meyer and Nelson, "Respecting What We Destroy: Reflection on Human Embryo Research," p. 18

15. Lanza et al., "The Ethical Validity of Using Nuclear Transfer in Human Transplantation," p. 3177.

16. Carol Tauer, "Embryo Research and Public Policy," *Journal of Medicine and Public Policy* 22, no. 5 (October 1997): 430.

17. Ronald Green also gives this argument. See Green, "Determining Moral Status," p. 22.

18. Meyer and Nelson, "Respecting What We Destroy: Reflection on Human Embryo Research," p. 18.

19. Jeffrey Spike, "Bush and Stem Cell Research: An Ethically Confused Policy," *American Journal of Bioethics* 2, no. 1 (2002): 45.

20. Carson Strong would endorse this principle. See Strong, "The Moral Status of Preembryos, Embryos, Fetuses, and Infants," p. 467.

21. The line of argument can be found in Bonnie Steinbuck, *Life before Birth: The Moral and Legal Status of Embryos and Fetuses* (New York: Oxford University Press, 1992).

22. Meyer and Nelson, "Respecting What We Destroy: Reflection on Human Embryo Research," p. 19.

23. Ibid., p. 18.

24. Mary Anne Warren, *Moral Status: Obligations to Persons and Other Living Things* (Oxford: Clarendon Press, 1997).

25. Eric Juengst and Michael Fossel, "The Ethics of Embryonic Stem Cells—Now and Forever, Cells without End," *JAMA* 284, no. 24 (December 27, 2000): 3180–84.

26. Mary Anne Warren, "On the Moral and Legal Status of Abortion," *Monist* 57, no. 1 (January 1973): 43–61. This essay has been widely reprinted.

27. See John Paul II, *Evangelium Vitae* (Boston: Daughters of St. Paul, 1995).

28. Lanza et al., "The Ethical Validity of Using Nuclear Transfer in Human Transplantation," p. 3177.

29. Green, "Determining Moral Status," pp. 20–30.

30. Strong, "The Moral Status of Preembryos, Embryos, Fetuses, and Infants," p. 460. This argument originally is due to Steven Buckle, "Arguing from Potential," *Bioethics* 2 (1988): 227–53.

17

Public Stem Cell Banks

Considerations of Justice in Stem Cell Research and Therapy

Ruth R. Faden, Liza Dawson, Alison S. Bateman-House, et al.

The possibility that stem cells can provide therapies for disease and illness has generated immense excitement on the part of both researchers and patients. This enthusiastic support for the notion of stem cell–based therapy is tempered by the fact that, at present, embryonic stem cells are considered technically superior to stem cells derived from other sources, such as umbilical cord blood or adult stem cells in the human body. Given this situation, policy decisions concerning stem cell research have become linked to the debate about the ethics of creation or destruction of embryos, leaving policymakers grappling with the seemingly intractable question of the moral significance of the human embryo. This debate will continue; however, it is inevitable that research into stem cell engineering will also continue. It seems equally inevitable that as this field of research develops, additional ethics and policy questions will arise.

The forthcoming transition in the focus of stem cell research

Reprinted from *Hastings Center Report* (November–December 2003): 13–27.

from basic science to the development of therapies raises important questions of justice. This transition is marked by increasing interest in establishing banks of stem cell lines, both to facilitate research and in anticipation of the eventual use of stem cell–derived transplants to treat such diseases as amyotrophic lateral sclerosis, Parkinson's, and diabetes.[1] The creation of stem cell banks raises questions about who stands to benefit from these banks and their research and therapeutic applications. First, there is a question about who, financially, will have access to stem cell–based therapies. Also, given that some nations have legislated against allowing the use of embryonic stem cells, there may be a question of who legally will have access to therapies derived from banked stem cell lines, particularly those of embryonic derivation.

A final issue, and the one we will discuss in [this chapter], is who *biologically* will have access to cell-based therapies. As we will show, the biological properties of stem cells themselves may make them less accessible to some potential recipients than to others, a situation we term the problem of *biological access*. Unless the problem of biological access is carefully addressed, an American stem cell bank may end up benefiting primarily white Americans, to the relative exclusion of the rest of the population. We must therefore ask which of all possible ways to structure an American stem cell bank is the most just.

REJECTION AND THE THEORETICAL SOLUTION OF AUTOLOGOUS GRAFTS

The future promise of cell engineering is the ability to control cells and their functions. In the interim, however, it seems likely that cell-based treatment for disease and injury will be orchestrated through the transplantation of stem cells or their products. As with more conventional types of transplants, immune rejection is a major potential problem. Immune rejection is the principle reason that a given stem cell-based therapy for a specific disorder might be biologically less available to one patient than to another. . . .

Some of the same limitations plague the second autologous strategy for solving the problem of rejection, that of using cells obtained directly from the patient herself through the identification and culturing of the patient's own adult stem cells.[2] Some claim that it is possible, at least in animal models, to derive adult stem cells that exhibit the same degree of developmental capacity as embryonic stem cells. From a public policy perspective, the adult sources alternative has great appeal, as it sidesteps altogether that difficult issue of embryo destruction. Whether adult sources are able to replace embryonic sources remains to be seen.[3] However, even if adult sources of stem cells are shown to be as robust as embryonic sources, using them to produce autologous stem cell–based therapies is problematic. At least for the near future, the laboratory procedures involved are extremely inefficient in generating sufficient cells. The isolation of adult stem cells yields very few cells, which are difficult to grow in culture. Like the cloning strategy, the adult stem cells strategy is both time consuming and expensive.

There may be some circumstances in which the time and expense required to prepare customized autologous therapies are justified. For example, a stem cell–based therapy that cures a young child of a burdensome condition, thereby saving the health care system a lifetime's worth of medical expenses while providing a profound benefit to the child, might justify the time and expense of creating an autologous therapy. For most conditions, however, the cost of customized autologous therapies would be prohibitive, even for wealthy nations. Moreover, for conditions such as stroke and injury, where treatments may need to be administered quickly in order to be maximally effective, it may never be possible to prepare autologous stem cell therapies from adult (or cloned) sources within the required time constraints. Although non-autologous transplants supported by immunosuppressive therapies could in theory be used to sustain stroke and trauma patients during the time required to prepare customized, autologous stem cell therapies, here, too, the costs are likely to be prohibitive. Therefore, adult sources are not much more likely than cloned sources to provide a complete solution to the rejection–biological access problem, at least for the foreseeable future.

ALTERNATIVE STRATEGIES FOR ADDRESSING REJECTION AND BIOLOGICAL ACCESS

For the remainder of this [chapter], we assume that there is no "autologous fix" to the problem of biological access, at least not until the capacity to engineer cells advances to the stage where in vivo manipulation of stem cells is commonplace. In the interim, human therapies derived from stem cells will probably involve transplantation of grafts from a genetically non-identical donor. Furthermore, we assume further that, even if interventions directed at organ systems such as the brain or liver are found to be relatively unproblematic, at least some stem cell–derived therapies will create immune rejection problems.

At present, there are three main options in dealing with the problem of immune rejection: immunosuppressive drugs, clinically induced tolerance, and HLA matching.[4] Often, two or more of these techniques are used in combination. A fourth technique, genetic modification of cell lines to reduce their capacity to provide an immune response, has no current application in clinical transplantation but perhaps could be used in future development of cell-based therapies. In theory, genetic modification could be used to create the equivalent of "universal" stem cells—cells that would not produce immune reactions in most patients.[5] From the perspective of justice, such a development would be ideal, since biological access for almost all persons would be guaranteed. Animal experiments suggest avenues for pursuing the universal stem cell strategy; however, the technical barriers to defeating the multiple defenses of the immune system are formidable.[6] It may well be years, if not decades, before such engineering will be successful.

The most widely used strategy to deal with immune rejection is immunosuppression. Immunosuppressive drugs began to be widely used in the 1980s, greatly increasing the viability of HLA-mismatched organ donations. In many cases, however, transplant recipients need continual immunosuppression with drugs in order to avoid either acute or chronic rejection, even when HLA matching is available. The risks of immunosuppressive therapy are well documented and often severe. They include nephrotoxicity, diabetic and vascular complications, and an increased risk of infections.[7]

Another strategy by which to avoid rejection is the induction of immunologic tolerance. Experiments with animals have shown that various methods of reducing host immune response and promoting acceptance of grafted tissue can reduce rejection and lessen the need for ongoing immunosuppression of the graft recipient.[8] However, clinical applications for humans are still in development and are at present relatively risky. A technique for inducing tolerance called mixed chimerism is particularly intriguing in light of the potential of a single stem cell line to generate different tissue types.[9] In mixed chimerism, the host immune system is temporarily suppressed, and donor bone marrow is introduced into the recipient and allowed to engraft prior to transplant of an organ from the same donor. If the technique is successful, the recipient develops a chimeric immune system consisting of her own immune cells and the new cells engrafted from the bone marrow. This chimerical immune system should be tolerant of new tissue (for example, a transplanted organ) from the same donor. At present, few patients have undergone this procedure, mainly due to the risks involved, the uncertainty of success, and the need for a living donor who can provide both bone marrow and an organ, such as a kidney. However, data from animal experiments are promising.[10] If the same stem cell line could be induced to produce hematopoietic cells for transplant and the tissue of interest for therapy, the mixed chimerism approach could be used to provide patients with cell-based therapies from stem cells without extensive immunosuppression or the need for HLA matching.[11]

The third strategy for avoiding immune rejection is the HLA matching; however, the importance of HLA matching in transplantation varies, depending on what tissue or organ is transplanted. For example, for bone marrow transplantation HLA matching is considered essential for a good clinical outcome,[12] while for liver transplantation, matching is not normally used.[13] The importance of HLA matching in transplantation also varies depending upon donor availability and disease severity. As mentioned above, while not considered optimal, mismatched transplants are performed, primarily if a match is not available.

The U.S. National Marrow Donor Program has compiled a registry of over four million donors, each of whom is typed for their HLA-A and B alleles, which are considered critical for matching. Due to the high degree of polymorphism in the relevant alleles, even with this enormous pool of donors only 50 to 60 percent of patients who need transplants can find a match.[14] Not only are the HLA alleles highly variable, but also different ethnic groups have different frequencies of specific alleles.[15] For example, the ten most common HLA-A alleles in white Americans are not the ten most common in African Americans, and vice versa.[16]

The transplant community has struggled with the issue of HLA distribution and its relation to immune rejection for years, in particular with regard to patients who are less likely to find a match due to their ethnic or racial background. This concern will extend from solid organ and bone marrow transplantation to stem cell transplants because stem cells bear the haplotype of the individual from whom the cell line was derived. Although the need for HLA matching of stem cell–derived therapies will likely vary depending on the tissue that is transplanted, matching will be critical to clinical success in at least some important therapeutic applications. As such, the disparities currently present in the field of transplantation are likely to be replicated in the emerging practice of stem cell transplantation, unless specifically guarded against.

We have addressed elsewhere the issue of whether the existing stem cell lines are suitable for use in human recipients.[17] Even assuming that current stem cell lines are appropriate for human use, they are woefully inadequate from the perspective of HLA matching. The situation in the United States is particularly acute. At present there is no publicly available information about the HLA types of the embryonic stem cell lines that are approved for federally funded research in the U.S. However, given the small number of lines[18] and the fact that these stem cells were derived from embryos created by in vitro fertilization for reproductive use, the diversity of HLA types among these lines is probably extremely limited. In the near term, the unlikelihood of haplotype diversity in available stem cell lines may significantly impede the efficiency and success of first human clinical trials. Looking ahead to therapeutic use, two concerns loom large: First, many patients will not be able to find a match and therefore will face more burdensome therapeutic regimens that are less likely to be successful. Second, some groups of people may be systematically disadvantaged if their ancestral/ethnic group was not well represented in the biological material that was initially used to derive stem cells, since their haplotypes are then less likely to be included in stem cell–based therapies.

A PUBLIC POLICY RESPONSE TO THE PROBLEM OF BIOLOGICAL ACCESS

We strongly recommend that all four of these strategies for dealing with immune rejection be actively pursued. Although the capacity to induce tolerance is currently in the earliest stages of clinical application, advances in this area hold great promise, not only for stem cell–based therapies but also for transplantation in general. Continued research into immunosuppression may lead to the development of next-generation drugs that have a reduced side-effect profile. The potential to develop a universal stem cell should be explored, although the scientific obstacles are formidable. In the near term, however, HLA matching, supplemented with immunosuppression as needed, remains the principle available approach to avoiding rejection.

HLA matching and transplantation raise serious questions of public policy and justice. In the American context, there have been many attempts to address one such issue: the relative unavailability of good matches for African American transplant recipients.[19] Public policy responses to this problem have generally been restricted to appeals to the African American community for donation and to strategies to increase overall donation. In the context of stem cell–based therapies, however, the availability of HLA types need not be constrained by the vagaries of organ donation. Although we are not currently able to produce solid organs or tissues for transplantation, we are able to create stem cell lines that can be used for research and, eventually, therapies. This means that it is within our power to construct a bank of stem cell lines that includes a wide spectrum of HLA types, specifically selected to satisfy considerations of justice. . . .

HOW SHOULD A THERAPEUTIC BANK BE DESIGNED?

We begin with a discussion of the issues that surround the creation of a bank constructed to serve therapeutic needs, then describe the distinct considerations that apply to the research context. It turns out that separate structures for research and therapy banks are desirable.

Ideally, a stem cell bank would include sufficient diversity to permit every potential recipient to receive a good match. Unfortunately, such a bank would require the creation and maintenance of a collection of enormous magnitude. As we have already noted, even with a registry of over four million donors, the U.S. Bone Marrow Donor Program provides matches for only 50 to 60 percent of those in need. The Bone Marrow Donor Program is a registry, providing searchable information on potential donors. Given the time needed to develop stem cell lines, we assume that a registry system would not be feasible for stem cell transplants. Rather, we propose a bank, where cell lines would be maintained and stored and samples distributed to clinicians as needed. We assume that constructing a stem cell bank with hundreds of thousands or even millions of stem cell lines is out of the question, if for no other reason than the huge financial cost of creating and maintaining so many stem cell lines. . . .

The central ethical challenge of this proposition is determining which combination of haplotypes to include in the limited bank of homozygous cell lines intended for therapeutic use. The first step toward addressing this challenge is an assessment of the options. We think there are three main strategies, each highlighting different considerations of justice, for the selection of cell lines to be included in a limited, homozygous public stem cell bank. A straightforward maximizing approach would seek to include those cell lines from which the most matches could be made. An egalitarian approach would give all individuals who it is feasible to include in the bank an equal chance at having their haplotype represented. What we call an ethnic representation strategy would select common haplotypes within each ancestral/ethnic group so that the members of any group would have the same chance of finding a match with the banked cell lines. We consider each of these strategies and argue that the last is the most defensible.

COVERAGE MAXIMIZING STRATEGY

The first strategy is to seek to include those homozygous cell lines that would allow the greatest percentage of the population to find a match in the bank. This strategy recognizes that not all cell lines are alike in terms of the number of people who might benefit from them. Some haplotypes are more common than others, and a limited bank can cover more people if it includes cell lines that possess the most common haplotypes. The obvious appeal of this strategy is that it provides for the largest number of potential beneficiaries of HLA matched stem cell–based treatments.

There are, however, two significant drawbacks to this approach. First, it ensures that persons with less common haplotypes could never benefit from the bank. One might reasonably be concerned about the fairness of such a strategy. Second, a bank composed of cell lines possessing the most common haplotypes in the United States would statistically favor white Americans simply because white Americans are the most populous group in the country.

The haplotypes that occur most commonly in white Americans overlap somewhat with the most common haplotypes of other American ancestral/ethnic groups, but significant diversity exists among the groups. . . . Even with the overlaps not all ancestral/ethnic groups share common haplotypes. The most common HLA-A/B haplotype within white Americans, A 0101 B 0801, is among the ten most common for African Americans, Hispanics, and Native Americans; however, this haplotype is not among the twenty-five most common for Asian Americans. . . .

Since white Americans are more numerous than America's other ancestral/ethnic groups, the inclusion of a haplotype found in a relatively small percentage of white Americans might extend coverage to more people than the inclusion of the haplotypes most common in another ancestral/ethnic group. For this reason, if a bank included homozygous lines with the fifty most common haplotypes in the United States, the de facto result would be a bank composed primarily of lines whose haplotypes are common to white Americans. While this strategy would lead to a higher number of matches than any other, the matches would be clustered within the Caucasian ancestral group, exacerbating the health discrepancies that currently exist between ethnic groups within the United States—discrepancies that track histories of oppression and social injustice.

EQUAL CHANCES STRATEGY

Once way of addressing the concern about fairness to individuals with less common haplotypes would be to give all haplotypes we can feasibly include in a bank, and thus all the individuals who have these haplotypes, an equal chance at being represented.[20] As a practical matter, it is effectively impossible to create homozygous stem cell lines for haplotypes that are sufficiently rare. It would be possible, however, to include in a bank many haplotypes that fall somewhere in between the rare and the common ones. The equal chances strategy seeks to promote fairness by giving all persons with haplotypes that can feasibly be represented in the bank the same chance at biological access to stem cell–based therapies. This could be accomplished by randomizing the process through which eligible haplotypes are selected for inclusion in the bank (for example, through some form of lottery in which all the relevant haplotypes are included).

While providing as many individuals as possible an equal chance of benefiting from the bank may accord with some basic intuitions about fairness, adopting the equal chances strategy has two real drawbacks. First, this strategy is not designed to address the problem of unequal access for members of different ancestral or ethnic groups. In practice, the equal chances approach might either alleviate or exacerbate these inequalities, depending on the outcome of the lottery. In either case, however, these results would be due to luck, not design, and might lead to even greater disparities between ancestral/ethnic groups than the coverage maximizing strategy. Some might argue that the ethnic inequalities that might result from an equal chances strategy are more morally acceptable than those that would result from a coverage maximizing strategy because the process that yielded them is fair. However, those who hold that there are strong independent moral reasons to prevent further disadvantages for historically oppressed groups will not be satisfied by a process that might have this result.

Second, while a lottery might, as a matter of luck, lead to the inclusion of the same set of haplotypes as the coverage maximizing strategy, the point of adopting the equal chances approach is to allow for other possibilities as well, including the possibility that most or all of the haplotypes included in the bank would be relatively uncommon. In this case, obviously, only a small number of persons would be able to benefit from the bank.

The problem that only a few might benefit from an equal chances bank becomes more acute when we consider which haplotypes we might reasonably exclude from the lottery on the grounds that they are so rare that they cannot feasibly be included. We must exclude any haplotypes that are so rare that it would be literally impossible to find donors of the gametes needed to create them. If we also excluded haplotypes on the grounds that it would not be impossible but merely extremely difficult and costly to find such donors, we would have to do so because we judged that the costs of including those haplotypes in the lottery would outweigh the benefits of doing so. But to exclude haplotypes from the lottery on this basis is inconsistent with the justification for the equal chances strategy.

That justification relies on two claims: first, that our attempts to benefit the greatest number of patients should be constrained by the requirements of justice, and second, that justice requires that we give those with uncommon haplotypes an equal chance of benefiting from the bank. If these assumptions are correct, then the fact that the equal chances strategy provides benefits to fewer patients than the coverage maximization strategy does not show that the equal chances strategy should not be adopted. By the same token, however, the fact that some haplotype is so uncommon that creating a homologous stem cell line with that haplotype would absorb most of the resources available to a bank cannot show that we would exclude it from the lottery. But if our lottery must include all such haplotypes, the number of uncommon haplotypes would be larger than one might have thought, and the probability that the bank would benefit only a very small number of people would be correspondingly greater.

We do not believe that justice requires the adoption of the equal chances strategy. In designing a bank to provide maximal coverage, we do not deprive those with uncommon haplotypes of a benefit to which they are antecedently entitled or ask them to make sacrifices from which they cannot expect to benefit. We are, instead, in a situation in which we must decide how best to allocate scarce resources. In other such situations we do not believe that the only fair way to make decisions is by lottery. For instance, those who allocate other scarce medical resources, such as ICU beds or organs, do not rely on lotteries to make their decisions, and we do not generally think that these practices are unfair. Depending on the context, allocation decisions take into account such factors as medical need or prognosis, even though this means that those who are less seriously ill or who are less likely to survive do not have an equal chance of securing the resources. Without some reason to regard the decision of which lines to include in a stem cell bank as different in kind from other decisions about the allocation of scarce resources to which no one is antecedently entitled, we should conclude that fairness does not require adoption of the equal chances strategy.

Still another reason for rejecting the equal chances strategy is that, at least for some, the primary justification for investing in stem cell research and the creation of a public bank is the advancement of human welfare, generally. From this standpoint, a process of creating the bank that yields very little benefit would be self-defeating. Whatever one thinks of the fairness of providing equal chances, we must not lose sight of the fact that in this instance we are seeking fairness within the context of advancing social welfare. If there is too little welfare or benefit, the putative fairness promoted through equal chances comes at too high a price.

ANCESTRAL/ETHNIC REPRESENTATION STRATEGY

Although we the authors do not all agree on this point,[21] most of us would prefer to select the most common haplotypes from each of the major ancestral/ethnic groups in the United States in order to make the bank useful to the same percentages of patients from each ethnic category. The strategy would be less efficient than the coverage maximizing strategy because it would take more cell lines to match the same number of patients overall. This is true for two reasons. First, the entire representation strategy holds that we should extend coverage to the same proportion of each ethnic group, even though a given percentage of a smaller group includes fewer people than the same percentage of a larger group. Second, different numbers of cell lines would be needed to cover the same percentage of different groups, due to the fact that some ethnic groups have more HLA diversity than others. For example, when matching HLA-A, B, and DR, in order to ensure that roughly 50 percent of all white Americans and 50 percent of all African Americans could receive a suitable match, between sixty and eighty cell lines would be needed. . . . Twenty homozygous cells lines would be sufficient to match 48.6% of white Americans, but only 28.7% of African Americans. In order to cover approximately 50 percent of each group, between twenty and thirty cell lines would be needed for white Americans and between forty and fifty for African Americans.

Matching for the DR alleles in addition to the A and B alleles decreases the likelihood of finding a match and increases the number of cell lines necessary to match a given percentage of a population. In some cases it may be reasonable to match only A and B, depending on the type of tissue transplanted and the likelihood of a good clinical outcome. When matching for HLA-A and B only, in order to cover 30 percent of each of the five ancestral/ethnic groups, approximately twenty-three cell lines for African Americans, twelve for white Americans, twenty-four for Hispanics, fourteen for Native Americans, and twelve for Asian Americans, would need to be established, for a total of eighty-five cell lines.[22] By contrast, if the eighty-five most common haplotypes in the overall U.S. population were included (irrespective of ethnicity), white Americans would make up most of the potential matches.

Since white Americans constitute 75 percent of the overall population,[23] the haplotypes most common in this group are the most common in the U.S. population overall. If the ten most common haplotypes among white Americans were chosen for the stem cell bank, only three would overlap with the ten most common haplotypes for African Americans and Native Americans; there would be four overlaps with Hispanics, and none with Asian Americans. Thus, such a bank would provide matches for a much higher proportion of white Americans than of any other ancestral/ethnic group.[24]

On our proposal, fewer patients would have access to stem cell therapies than would otherwise be the case. We do not take lightly the idea of designing a bank in such a way that fewer patients will be able to benefit from it. Nonetheless, we believe that the ethnic representation strategy should be adopted. In the United States, ancestral/ethnic groups other than white Americans are the only groups of persons that share two traits: first, they would be systematically underrepresented in a bank constructed according to the coverage maximization strategy, and second, they have endured a history of discrimination within American society. The coverage maximization strategy would both mimic this discrimination and exacerbate its effects, which in our view argues against its adoption.

As members of societies that have a history of ethnic discrimination, we have an obligation to reduce ethnic disparities in life expectancy and other indicators of health. Insofar as these disparities are understood as present injustices, at the very least, public policy should not be formulated in ways that make them worse.[25] Insofar as they are the result of past injustices, as members of the society that produced them, we have an affirmative obligation to take steps to ameliorate them. For these reasons, it would be wrong to adopt policies that exacerbate the effects of discrimination, even if the factors that would serve to widen the disparity—for example, a higher rate of polymorphisms in one group as compared to another—are themselves unrelated to any historical or current social injustices.

Moreover, providing equal ethnic representation in a stem cell bank would prevent the expressive harm that would result from unequal representation. If we followed the coverage maximizing strategy, the resulting stem cell bank would ensure access to stem cell–based therapies for a much greater percentage of white Americans than other groups. For example, if the twenty-five most common haplotypes among all Americans were selected, due to the fact that white Americans are the most numerous group, all twenty-five haplotypes would be those common to white Americans. Thus, in the pool of twenty-five cell lines, approximately 40 percent of white Americans could find a match, while 7.8 percent of African Americans could be matched by the pool of cell lines, 19 percent of Hispanics, 21 percent of Native Americans, and 3.6 percent of Asian Americans. The justification for adopting this strategy is based solely on a commitment to maximizing medical benefits, without regard to the implications for the different ethnic groups. Indeed, had the population genetics worked out differently, the coverage maximizing strategy could have affected ethnic groups quite differently. Nevertheless, if a bank made the benefits of stem cell therapy available almost exclusively to white Americans, members of minority ancestral/ethnic groups might well wonder whether their interests had been taken seriously by those who decided which lines to include. Given the history of American race relations, and of the medical profession's treatment of non-white Americans, this concern cannot be dismissed as unreasonable.[26] The need to avoid giving some persons reasonable grounds for concern about whether they are regarded as full and equal citizens whose interests are taken seriously, especially when those concerns have often been well founded, is a further reason to reject the coverage maximizing strategy.

While the ethnic representation approach is not maximally efficient, it does ensure that the greatest amount of benefits is produced consistent with an expression of respect for the fundamental equality of members of at least the major ancestral/ethnic groups in the United States.[27] Given the country's history of oppression of a number of minority groups and the continued fragility of race relations, a policy that allowed further privileging of white Americans over other groups would signal a failure to acknowledge the equal worth of persons of all ethnic groups.

A STEM CELL BANK FOR CLINICAL RESEARCH

We now turn to the question of how a research bank should be constructed. The goals of clinical research are distinct from the goals of clinical medicine, and so too are the relevant moral considerations. Everyone has an interest in research yielding its results as efficiently as possible and thus everyone has an interest in investigators being able to find appropriate human subjects quickly and easily. In contexts where HLA matching is thought to be important, it will be much easier to find eligible research subjects if the stem cell line from which the intervention is developed has a common haplotype. Thus, in a research bank, as opposed to a therapeutic bank, the arguments favoring the equal chances strategy have no force. The argument in favor of the ethnic representation strategy may also seem less persuasive, since the primary concern is to establish quickly whether a particular experimental treatment is indeed "safe and effective" and thus worth distributing to all.

We agree that a research bank should be designed to fit the needs of the research enterprise and thus that it should be composed primarily of homozygous stem cell lines from the most common haplotypes in the American population. However, there is a powerful argument for including at least several homozygous lines that are common in particular ancestral/ethnic groups. Without such lines, it is possible that researchers will be both less able and less likely to pursue the promise of stem cell science for diseases that occur disproportionately or present differently in different ethnic groups. If this were to occur, then it would not be possible for all to benefit fairly from society's investment in stem cell research. Assuming that there are good arguments for keeping the number of lines in a research bank to a minimum, a research bank of homozygous stem cell lines could likely function effectively with as few as fourteen lines—the six most common haplotypes of the population, which would match approximately 25 percent of all Americans (most of whom would be white Americans), as well as the two most common haplotypes in African Americans, Hispanics, Native Americans, and Asian Americans, which would match between 5 and 10 percent of the population in each of these ethnic groups.

GLOBAL JUSTICE

In this [chapter], we focus on research and therapy banks for the United States, and our analysis of how to construct these banks justly is specific to the American context. In stem cell banks designed for other countries or for multinational banks, considerations of justice may well be specified differently and thus different patterns of haplotypes may be required.

A particularly important worry from the perspective of justice is how fairly to accommodate the world's population as stem cell medicine progresses. Data from the population genetics literature indicates that populations in different regions are likely to have significantly different HLA frequencies—both different from each other and different from the U.S. population—thus potentially confounding efforts to make therapies widely available on a global scale. For example, sub-Saharan African populations exhibit the highest degree of genetic diversity globally,[28] and this diversity is not well represented in groups in other world regions. Economic considerations would clearly come into play for countries in the global South, whose health care and health research budgets are already severely constrained—but again, this topic merits a separate analysis and is not the focus of our efforts. We assume that relatively rich countries will develop stem cell–based therapies and that eventually these products will be made financially available to those in poorer countries. To achieve biological access on a worldwide level, concerted effort and collaboration will be needed among developed nations pursuing stem cell–based therapies in order to consider genetic diversity in sufficiently broad terms to meet the needs of patients in resource-poor, as well as resource-rich, countries.

MORAL AND POLITICAL CHALLENGES

There are several significant challenges to creating patterned stem cell banks in the manner we have proposed. Assuming for the time being that the cell lines will be derived from embryonic sources, the first challenge will be the solicitation of gametes. Many people will need to be HLA typed in order to identify donors who have the desired haplotypes. Female donors will have to undergo the burdensome process of ovarian hyperstimulation and oocyte retrieval, the risks and discomforts of which are not trivial. The acceptability of these risks turns in part on how they compare to the risks and discomforts of donating bone marrow or of being a living kidney or liver donor. Like these other transplantation donors, gamete donors to a stem cell bank should not be paid, thereby sharply distinguishing the banks from the practices of infertility programs. The burdens of ensuring a just system of access to stem cell therapies will fall disproportionately on women relative to men (for whom gamete donation is, by comparison, inconsequential). Whether women will be willing to become egg donors in the absence of financial compensation is unclear, although based on experience with the donation of bone marrow, kidneys, and livers, many people appear willing to assume medical burdens for the benefit of others. It is also possible that laboratory procedures will be developed to drive differentiation of human embryonic stem cells into oocytes,[29] obviating the need for egg donations from individual women. This technology has not yet been fully worked out, and thus cannot yet be counted on for establishing a stem cell repository.

A related challenge will be securing sufficient gamete donations from minority populations and, in particular, from African Americans. The whole point of the ethnic representation strategy is to ensure that minorities are not systematically disadvantaged in access to stem cell therapies. At the same time, however, the African American community is distrustful of the medical and scientific establishment. This distrust manifests itself in many ways, including reluctance to consent to organ donation and reluctance to participate in medical research. Since constructing the banks as proposed will be impossible if African Americans and other minority groups do not participate in it, securing their trust and commitment will be essential.

The most obvious, and most formidable, challenge to creating stem cell banks in the United States is the widespread disagreement about the moral status of early human life. It is certain that a significant portion of the population will be opposed to the creation of such banks solely because they necessitate the creation and the destruction of embryos. It may be difficult for politicians or governmental entities to support the idea of a patterned stem cell bank because of the amount of controversy surrounding this very contentious issue.

At least in the near term, creating the desired pattern of homozygous cell lines will require deriving lines from new embryonic sources. Developing a just system of access to the benefits of stem cell therapies would thus appear to require the instrumental creation and destruction of embryonic life.[30] Therefore, we believe that it is morally desirable to delay creation of the therapy bank until there is solid evidence from early clinical trials that stem cell–based therapies will work. In the interim, we should examine the progress that is being made with non-embryonic sources of stem cells and with immunosuppression and tolerance-inducing techniques. If any of these approaches are significantly advanced by the time stem cell therapies are approaching clinical utility, it might render a therapy bank created through the destruction of embryos unnecessary.

At the same time, however, it is essential to establish a research stem cell bank in order to justly and safely proceed with human clinical investigation. Several avenues of research in stem cell science are approaching first human experiments. Elsewhere, we argue that the embryonic stem cell lines currently approved for federal funding are not appropriate for use in human beings.[31] Unless adult sources of stem cells can, in the very near term, be determined to produce robust stem cell lines, it is likely that the transition from the laboratory to clinical investigation will require the destruction of additional human embryos. A patterned research bank constructed of homozygous lines of common haplotypes may actually minimize this use of embryos. Possibly as few as fourteen lines would provide a sufficiently broad base for clinical research, including the investigation of applications of particular interest to minority communities.

Another challenge will be identifying a structure for the research bank that will allow it to function as a public good and thus to fulfill its social purpose. A complicated web of proprietary interests has made it very difficult for researchers to effectively use existing stem cell lines. It is unclear whether a research bank could be constructed that could avoid this morass, particularly if it is not established or regulated by the federal government. Since federal involvement in a research bank is unlikely, funding will need to come from the private sector. Philanthropic support would be more likely to ensure that the bank operates as a true public good than would a consortium of commercial interests. By the time a therapy bank needs to be constructed, government involvement may be possible. For example, public values may shift, should the clinical utility of embryonic stem cell lines be established. Alternatively, non-embryonic cells might become reliable sources of stem cell lines, allowing the therapy bank to be constructed without the use of embryos.

Although there is encouraging progress in research on adult sources, we are not optimistic that there will be a technical fix for the moral and public policy quandaries posed here. It seems most likely to us that evidence of therapeutic values will be at hand before alternatives to embryonic sources will be found to be practical. Although we strongly support continued research into better immunosuppressive therapies and tolerance induction and believe that advances will be made in this area, it also seems unlikely to us that they will render the clinical advances of HLA matching moot. Thus, we believe that society may well have to choose what it values more—ensuring that all benefit fairly from advances in stem cell science or protecting embryonic human life. If society decides to create a therapy bank, then every effort should be made to coordinate with similar efforts in other countries, in order to minimize the number of embryos that must be destroyed. The United Kingdom recently announced that it has already embarked on the creation of a stem cell bank of its own.[32] It is not known at this writing whether the U.K. bank is being designed to address considerations of justice. It is also not clear what kind of HLA distribution is represented in the U.K. bank or whether immunologic matching would be possible for some proportion of the U.S. population.

Current and future policies concerning scientific research need to be responsive to concerns about equitable biological access addressed in this [chapter]. The existing human embryonic stem cell lines in the United States on which federally funded research is allowed will be insufficient to meet this goal. Federal restriction on stem cell research will need to be reevaluated, along with policies regarding funding priorities, patent protections, and incentives to the research community in order to ensure that justice concerns are adequately addressed as scientific research progresses. Although the process will be controversial, the need for equitable biological access to new therapies must be balanced with respect for early human life. Thoughtful discussion among scientists, policymakers, and the public about these challenging issues will help ensure that new therapies are developed fairly and responsibly.

ACKNOWLEDGMENTS

We gratefully acknowledge the contributions of Mary Carrington, James Lautenberger, and Ted Gooley to this [chapter,] a product of a grant funded by the Greenwall Foundation: "Ethics and Cell Engineering: The Next Generation."

AUTHORS

Ruth R. Faden, Ph.D., Phoebe R. Berman Bioethics Institute, Johns Hopkins University

Liza Dawson, Ph.D., Phoebe R. Berman Bioethics Institute, Johns Hopkins University

Alison S. Bateman-House, M.A., Phoebe R. Berman Bioethics Institute, Johns Hopkins University

Dawn Mueller Agnew, Ph.D., Developmental Genetics Laboratory, Johns Hopkins University School of Medicine

Hilary Bok, Ph.D., Department of Philosophy, Johns Hopkins University; Phoebe R. Berman Bioethics Institute, Johns Hopkins University

Dan W. Brock, Ph.D., Department of Clinical Bioethics, National Institutes of Health

Aravinda Chakravarti, Ph.D., McKusick-Nathans Institute of Genetic Medicine, Johns Hopkins University

Xiao-jiang Gao, Ph.D., SAIC, National Cancer Institute, Frederick, Maryland.

Mark Greene, Ph.D., Greenwall Fellowship Program in Bioethics and Health Policy, Johns Hopkins and Georgetown Universities

Patricia A. King, J.D., Georgetown University Law Center
Stephen J. O'Brien, Ph.D., American Genetics Association, Buckeystown, Maryland
David H. Sachs, M.D., Transplantation Biology Research Center, Massachusetts General Hospital, Harvard Medical School
Kathryn E. Schill, M.A., Phoebe R. Berman Bioethics Institute, Johns Hopkins University
Andrew Siegel, J.D., Ph.D., Department of Gynecology and Obstetrics, Johns Hopkins University School of Medicine; Phoebe R. Berman Bioethics Institute, Johns Hopkins University
Davor Solter, M.D., Ph.D., Department of Developmental Biology, Max Planck Institute of Immunobiology, Freiburg, Germany
Sonia M. Suter, J.D., George Washington University Law School
Catherine M. Verfaillie, M.D., Stem Cell Institute, University of Minnesota
LeRoy B. Walters, Ph.D., Kennedy Institute of Ethics, Georgetown University
John D. Gearhart, Ph.D., Institute for Cell Engineering, Johns Hopkins University School of Medicine; Department of Gynecology and Obstetrics, Johns Hopkins University School of Medicine

NOTES

1. See G. Vogel, "Pioneering Stem Cell Bank Will Soon Be Open for Deposits," *Science* 297 (2002): 1784; "China Approves Stem Cell Bank," BBC News World Edition, December 11, 2002.

2. Y. Jiang et al., "Pluripotency of MEsenchymal Stem Cells Derived from Adult Marrow," *Nature* 418 (2002): 41–49.

3. C. M. Verfaillie, "Adult Stem Cells: Assessing the Case for Pluripotency," *Trends in Cell Biology* 12 (2002): 502–508.

4. See J. A. Bradley, E. M. Bolton, and R. A. Pedersen, "Stem Cell Medicine Encounters the Immune System," *Nature Reviews: Immunology* 2 (2002): 859–71.

5. J. S. Odorico, D. S. Kaufman, and J. A. Thomson, "Multilineage Differentiation from Human Embryonic Stem Cell Lines," *Stem Cells* 19 (2001): 193–204.

6. N. S. Liao et al., "MHC Class I Deficiency: Susceptibility to Natural Killer (NK) Cells and Impaired NK Activity," *Science* 253 (1991): 199–202.

7. J. G. O'Grady et al., "Tacrolimus versus Microemulsified Ciclosporin in Liver Transplantation: The TMC Randomised Controlled Trial,"

Lancet 360 (2002): 1119–25; D. F. Schafer and M. F. Sorrell, "Optimising Immunosuppression," *Lancet* 360 (2002): 1114–15.

8. X. Yu, P. Carpenter, and C. Anasetti, "Advances in Transplantation Tolerance," *Lancet* 357 (2001): 1959–63.

9. M. Sykes, "Mixed Chimerism and Transplant Tolerance," *Immunity* 14 (April 2001): 417–24; D. H. Sachs, "Mixed Chimerism as an Approach to Transplantation Tolerance," *Clinical Immunology* 95 (2000): S63–S68.

10. F. Fändrich et al., "Preimplantation-Stage Stem Cells Induce Long-Term Allogenic Graft Acceptance without Supplementary Host Conditioning," *Nature Medicine* 8 (2002): 171–78.

11. Because of the time required for the patient to develop immunologic tolerance, this technique would not be useful for patients that suffer a sudden, acute injury and require immediate treatment.

12. G. Dini et al., "Unrelated Donor Marrow Transplantation for Chronic Myelogenous Leukaemia," *British Journal of Haematology* 102 (1998): 544–52; E. B. Edwards, L. E. Bennett, and J. M. Cecka, "Effect of HLA Matching on the Relative Risk of Mortality for Kidney Recipients: A Comparison of the Mortality Risk after Transplant to the Mortality Risk of Remaining on the Waiting List," *Transplantation* 64 (1997): 1274–77.

13. United Network for Organ Sharing, 2003, http://www.unos.org/.

14. P. G. Beatty et al., "Probability of Finding HLA-Mismatched Related or Unrelated Marrow or Cord Blood Donors," *Human Immunology* 61 (2000): 834–40.

15. H. B. Oh et al., "Probability of Finding HLA-Matched Unrelated Marrow Donors for Koreans and Japanese from the Korean and Japan Marrow Donor Programs," *Tissue Antigens* 53 (1999): 347–49; M. A. Blanco-Gelaz et al., "Genetic Variability, Molecular Evolution, and Geographic Diversity of HLA-B27," *Human Immunology* 62 (2001): 1042–50; S. T. Cox et al., "Further Diversity at HLA-A and -B Loci Identified in Afro-Caribbean Potential Bone Marrow Donors," *Tissue Antigens* 57 (2001): 70–72; R. Rajalingam et al., "Distinctive KIR and HLA Diversity in a Panel of North Indian Hindus," *Immunogenetics* 53 (2002): 1009–19; P. G. Beatty, M. Mori, and E. Milford, "Impact of Racial Genetic Polymorphism on the Probability of Finding an HLA-Matched Donor," *Transplantation* 60 (1995): 778–83.

16. K. Cao et al., "Analysis of the Frequencies of HLA–A, B, and C Alleles and Haplotypes in the Five Major Ethnic Groups of the United States Reveals High Levels of Diversity in these Loci and Contrasting Distribution Patterns in these Populations," *Human Immunology* 62 (2001): 1009–30.

17. L. Dawson et al., "Safety Issues in Cell-Based Intervention Trials," *Fertility and Sterility* 80 (2003): 1077–85.

18. As of October 14, 2003, twelve lines were available for shipping. NIH Human Embryonic Stem Cell Registry, http://stemcells.nih.gov/registry/index.asp.

19. B. Barger et al., "The Impact of the UNOS Mandatory Sharing Policy on Recipients of the Black and White Races—Experience at a Single Renal Transplant Center," *Transplantation* 53 (1992): 770–74; D. E. Butkus, E. F. Meydrech, and S. S. Raju, "Racial Differences in the Survival of Cadaveric Renal Allografts. Overriding Effects of HLA Matching and Socioeconomic Factors," *NEJM* 327 (1992): 840–45; M. H. Park, D. E. Tolman, and P. M. Kimball, "Disproportionate HLA Matching May Contribute to Racial Disparity in Patient Survival Following Cardiac Transplantation," *Clinical Transplantation* 10 (1996): 625–28.

20. Nicholas Rescher discusses a variation on this strategy in his article, "The Allocation of Exotic Medical Lifesaving Therapy," *Ethics* 79 (1969): 173–86.

21. LeRoy Walters, Dan Brock, and Mark Greene do not endorse this position.

22. In fact, somewhat fewer than eighty-five cell lines would be needed overall because of the partial overlap of common haplotypes among ancestral/ethnic groups. The point we are emphasizing here is that different numbers of cell lines would be needed to cover the same percentage of each group.

23. "Overview of Race and Hispanic Origin: U.S. Census Brief," U.S. Census Bureau, issued March 2001, http://www.census.gov/prod/2001pubs/c2kbr01-1.pdf.

24. It is important to emphasize again two points about using ethnic categories to describe biological phenomena such as HLA diversity. First, the socially defined categories lack precision in terms of ancestry. For example, a person who self-identifies as Hispanic may have ancestors from one or more of four major world regions: North America, South America, Africa, and Europe. Second, a related point: all population subgroups in the United States have some degree of admixture, that is, mixing of groups with different ancestral origins, and this is particularly evident in African American and Hispanic populations. Because of the complex history of human migration, population growth, and mixing, there is no clear way to predict what genetic similarities might exist in different populations. Therefore an empirical approach is needed.

25. Increases in health disparities do not necessarily result in injustice. If, for example, allowing greater health disparities made the worse-off groups better off than they would otherwise be, such disparities would arguably be consistent with principles of justice. Similarly, if a medical advance can only benefit some, but not all, any subsequent increase in health disparities is not necessarily unjust. The presumption against widening the gap is thus in principle a defensible one. However, we do not see any considerations that would defeat the presumption in favor of reducing health disparities in the present context.

26. That this concern is reasonable is an important part of this argument. We are not arguing that a stem cell bank should be designed to avoid any possible suspicion that those who decided which cell lines to include discounted some people's interests. This demand would be unreasonable: there is no policy so obviously noble that no one could conceivably misinterpret the motives behind its adoption. However, the creation of a stem cell bank that covers most white Americans and few nonwhite Americans is not just a policy that someone might possibly misinterpret as reflecting a complete disregard for the interests of nonwhite Americans. It is a policy that plainly invites that interpretation, especially in light of America's history of discrimination against members of minority ancestral/ethnic groups. See V. N. Gamble, "Under the Shadow of Tuskegee: African Americans and Health Care," *American Journal of Public Health* 87 (1997): 1773–78. Moreover, it would, in our judgment, be difficult to allay the suspicion that those who chose to design a stem cell bank in such a way that virtually no nonwhite Americans could benefit from it had failed to take the interests of nonwhite Americans seriously.

27. Ideally, the bank should include stem cell lines that are representative of all ethnic groups that have distinctive HLA patterns, including Native Americans. Whether this will be possible is unclear, both in terms of practical feasibility and in terms of resources.

28. A. Sanchez-Mazas, "African Diversity from the HLA Point of View: Influence of Genetic Drift, Geography, Linguistics, and Natural Selection," *Human Immunology* 62 (2001): 937–48.

29. K. Hübner et al., "Derivation of Oocytes from Mouse Embryonic Stem Cells," *Science* 300 (2003): 1251–56.

30. Although many countries currently prohibit the creation of a bank that necessitates the creation and destruction of human embryos, there are several countries in which such activity would be legally permissible. Furthermore, those countries that have criminalized this process might be more apt to revise their laws once the clinical applications of stem cell–based therapies leave the realm of speculation and enter that of observable reality.

31. L. Dawson et al., "Safety Issues in Cell-Based Intervention Trials," *Fertility and Sterility* 80 (2003): 1077–85.

32. National Institute for Biological Standards and Control, the UK Stem Cell Bank, January 6, 2003 [online], http://www.nibsc.ac.uk/divisions/cbi/stem-cell.html.

Part 4
Religious Issues

Editors' Introduction

We start this section with two pieces intended to show just how complicated this whole issue truly is. The first piece, on the rights and demands of the sick, makes one wonder if one group in society ever has the moral authority to impose its religiously based beliefs on other groups in society. The second, by Ted Peters, shows that there are issues of beneficence and nonmalfeasance at work in the ethical analysis that go beyond the traditional abortion debate when it comes to stem cell issues. This might lead one to think that some religious-minded people could accept that the moral status of an embryo is not relevant when that embryo would be destroyed come what may. Aline Kalbian's paper shows, on both biblical and philosophical grounds, that the Catholic Church is opposed absolutely to any work that tampers with the integrity of the unborn. There seems to be no middle ground for the Catholic Church when it comes to the sanctity of life for the unborn.

In the last article in this section we have a thorough work by LeRoy Walters. In it, he tackles stem cell research and the religious traditions in four regions of the world. He shows how religious traditions have substantially influenced the public policy of each region, and in doing so provides the basic views of each religion on the issue of stem cell research, which is a nice contrast to the detailed analysis presented by Kalbian's chapter. It is an ideal place to begin a comparative inquiry of the views of different religious traditions with respect to stem cell research.

QUESTIONS TO THINK ABOUT

(1) How much influence should religion have on public policy on the local, federal, and international stages with respect to stem cell research?
(2) How much weight should be placed on the comments of patients in this debate since they are the hopeful beneficiaries of new treatments derived from the use of stem cells?
(3) If two religious traditions give recommendations that are different with respect to stem cell research, how should one decide which view is right?
(4) Does religious tolerance require believers to be tolerant of secular values with respect to science?
(5) What other than religion could give rise to beliefs in how to behave morally?

18

Patients' Voices

The Powerful Sound in the Stem Cell Debate

Daniel Perry

Millions of patients may benefit from the applications of stem cell research, although there is disagreement about whether public funds should be used to develop the science. Patients have been key to winning political support. Acting as advocates, they have contended that public investment will speed the research and bring accountability to biomedical technology. A political dispute about the new research, which holds the potential for cures to devastating diseases and to foster healthy aging, shows the need to respect public sensibilities and to court public approval, as well as the importance of involving patients in debates where the methods of biomedical discoveries and ethical beliefs collide.

The achievement of isolating and growing cultures of self-renewing human pluripotent stem cells has set off waves of optimism among both researchers and the lay public.[1] The promise is tangible for effective new approaches to incurable diseases and underlying biological processes.[2] As shown in Table 1, over 100 mil-

---

lion Americans suffer from illnesses that might be alleviated by cell transplantation technologies that use pluripotent stem cells. Yet some representatives in Congress and some of the lay public, as well as religious groups such as the National Conference of Catholic Bishops, oppose putting public funds behind the technology. They say that stem cell research belongs under a federal ban that currently prohibits federal funding of embryo research.[3]

PATIENTS FOR RESEARCH

In 1999, a coalition of three dozen national nonprofit patient organizations, the Patients' Coalition for Urgent Research (CURe), emerged to argue for public funding of human embryonic stem cell research under guidelines of the National Institutes of Health (NIH). This would achieve two goals: (1) participation by the broadest number of scientists under established peer-review mechanisms, thus rewarding the most promising research and speeding progress, and (2) public accountability and guidelines developed through processes that allow for public comment on an area of science that has raised ethical concerns.[4]

Why a patients' coalition? As taxpayers, patients and their family members are entitled to expect their government to make the most of a substantial public investment in biomedical research through the NIH and other agencies. And as the bearers of the ultimate burden when medicine cannot relieve their suffering, patients are the most compelling witnesses to the value of research that quite literally can save their lives.

In general, the patients and their advocates who are active for CURe display tempered optimism when it comes to appraising the chances of anyone's health benefiting soon from applications of stem cell research. Furthermore, broad views on the ethics and appropriateness of the technology have been expressed by those in CURe. For example, they believe in the principles of informed consent and free choice. Stem cell research must not lead to an underground black market in "spare" embryos for research. In addition, women and men, as individuals or as couples, should not be paid to produce embryos for research purposes.

The stories of patients and family members have fostered bipartisanship on Capitol Hill and have effectively complemented other activities such as the stance voiced by leading theologians from four major faiths—Roman Catholicism, Protestantism, Judaism, and Islam—who, noting the calls of their religions for compassion for the sick, wrote a joint letter to Congress urging federal involvement.[5]

THE BROADER STAKES

The promise of human pluripotent stem cell research increases the likelihood that vastly more people will experience healthy and productive aging. Age-related disease costs billions of dollars and burdens millions physically and financially.[6] The additional costs in medical and long-term care that are incurred annually in the United States because its Medicare recipients lose their functional independence are calculated at $26 billion.[7]

One can imagine the cost twenty years from now in the United States alone, when the population over age sixty-five is expected to double and the number of Americans over age eighty-five is projected to quadruple.[8] Unless bioscience engenders and receives

Table 1. Persons in the United States Affected by Diseases That May Be Helped by Human Pluripotent Stem Cell Research. Data Are from the Patients' Coalition for Urgent Research, Washington, D.C.

Condition	Number of persons affected
Cardiovascular diseases	58 million
Autoimmune diseases	30 million
Diabetes	16 million
Osteoporosis	10 million
Cancer	8.2 million
Alzheimer's disease	4 million
Parkinson's disease	1.5 million
Burns (severe)	0.3 million
Spinal cord injuries	0.25 million
Birth defects	150,000 (per year)
Total	128.4 million

broad popular support, in the future, nations like the United States, which have a rapidly increasing aging population, will more than likely struggle with a much greater health care burden. This is why it is so important to respect public sensibilities and to court public approval fervently, even though it is also likely that the next discoveries will, too, collide with the ethical and religious beliefs of some.

In the stem cell debate, patients have stepped forward to help draw the line between science in service to the community and science for lesser motives. Sadly, some of their most compelling stories will be silenced before long by the progression of their diseases. It surely behooves us to remember their contributions and to engage their successors, who will continue to put a human face on the promise of biomedical research.

NOTES

1. J. A. Thomson et al., "Embryonic Stem Cell Lines Derived from Human Blastocycsts," *Science* 282 (1998): 1145; M. J. Shamblott et al., "Derivation of Pluripotent Stem Cells from Cultured Human Primordial Germ Cell," *Proceedings of the National Academy of Sciences of the United States of America* 95 (1998): 13726.

2. American Association for the Advancement of Science and Institute for Civil Society, *Stem Cell Research and Applications: Monitoring the Frontiers of Biomedical Research*, November 1999.

3. Rep. J. Dickey (R-Ark), "No Such Thing as Spare Embryos," *Roll Call* (June 3, 1999), p. 4. R. M. Doerflinger, testimony on behalf of the Committee for Pro-Life Activities of the National Conference of Catholic Bishops before the Senate Appropriations Subcommittee on Labor, Health, and Education hearing on legal status of embryonic stem cell research (Senate Hearing 105-939), January 26, 1999 [online], frwebgate.access.gpo.gov/cgibin/getdoc.cgi?dbname5105_senate_hearings&docid5f:54769.wais.

4. Goals adopted by Patients' CURe, Washington, D.C., May 20, 1999.

5. "Theologians from Four Major Faiths Express Support for Federal Funding of Stem Cell Research," press release from Patients' CURe, Washington, D.C. (October 14, 1999).

6. Alliance for Aging Research, *A Call for Action: How the 106th Congress Can Achieve Health and Independence for Older Americans Through Research*, 1999.

7. Alliance for Aging Research, *Independence for Older Americans: An Investment for Our Nation's Future. A Report by the Alliance for Aging Research*, 1999.

8. Ibid.

19
The Stem Cell Controversy

Ted Peters

This past summer, on August 9, 2001, President George Bush filled our television screens with his executive countenance to announce his long-awaited decision regarding federal support for stem cell research. Recall, he decided to support such research on existing stem cell lines; but he would not support the development of new lines from the destruction of embryos, even embryos derived from otherwise surplus fertilized ova in reproductive technology clinics.

This is an issue to which I have devoted extensive attention. I was appointed to the Ethics Advisory Board of the Geron Corporation a couple of months prior to the first isolation of embryonic stem cells (ES cells) by Jamie Thomson at the University of Wisconsin in August 1998 and the isolation of embryonic germ cells (EG cells) by John Gearhart at Johns Hopkins University a month later. Both of these discoveries were paid for by Geron. My colleagues and I have scrupulously followed each scientific breakthrough and each turn in the ethical debate ever since. Up until this point, however, most of

Reprinted from *Dialog: A Journal of Theology* 40, no. 4 (Winter 2001).

my reflection on this subject has been for the benefit of Geron, which is my role as an internal ethics advisor. I have not been asked to speak publicly on behalf of Geron; nor do I consider it my job to provide Geron with public justification for its activities.

Yet, more recently I, along with one of my colleagues, Gaymon Bennett, have begun to feel we have the resources to offer an informed ethical assessment of what is at stake in the now worldwide stem cell controversy. Our telephone at the Center for Theology and the Natural Sciences in Berkeley is ringing off the hook—actually newer telephones no longer have hooks, do they?—with calls from the media. So, Gaymon and I have put our heads together. Perhaps *dialog* readers might like to know what we think about this. Here goes.

THE POPE AND THE PRESIDENT

Recall the course of events. Over a period of months during Spring 2001 President Bush called to Washington a number of experts and stakeholders, including Geron scientists and Roman Catholic spokespersons. On the eve of his televised address, President Bush visited Rome. In a public statement connected with this visit, Pope John Paul II stated that support of embryonic stem cell research evidences a trajectory of moral corruption, a coarsening of consciousness that finds its birth in the assault on innocent human life in the womb and its maturity in acquiescence before such evils as "infanticide" and euthanasia. Support of embryonic stem cell research, according to the pontif, bespeaks moral desensitization. Opposition to stem cell research is an oblique appeal to nonmaleficence: do no harm to early embryos!

Let's note a couple items here. First, this is not merely a matter of science and public policy; it is a religious issue. Second, the single most influential Christian leader in the world villifies his opponents on the issue, by suggesting that scientists engaged in stem cell research are no better than those who commit infanticide.

BEYOND THE ABORTION DEBATE

The accusation is not new. It is a rehearsal of rhetoric made familiar in the abortion debate. In fact, opponents of embryonic stem cell research have cast the debate surrounding this research as nothing but the next chapter in the abortion controversy.

With all due respect to the Holy Father, however, the ethical issues involved with this research are far too complex to be reduced to such a simple assessment. Portraying the stem cell debate as the abortion controversy is at best intellectually misleading, at worst ethically negligent. Claiming a moral monopoly, opponents of embryonic stem cell research appear to overlook fundamental theological and ethical principles that might inspire support of this research.

The moral poignancy of the stem cell debate lies not in its proximity to the abortion controversy, but in the wider question of the potential good promised by this research. Stem cell research is morally significant first because of its promise of healing.

Without a doubt the potential future advance in human health due to stem cell research is possibly the most stunning news in the medical sciences in decades. With the help of stem cells, the body may be able to regenerate itself. Stem cells are the precursors to virtually every cell in the body. When harvested and invested with the genetic code of the person suffering from tissue degeneration, stem cells may rejuvenate failing tissues: organs, nerves, spines, and brain tissue. Teaching the body to heal itself, stem cell therapies, if predictions are accurate, may ease the suffering of millions afflicted with such debilitating diseases as Parkinsons, heart and liver failure, juvenile diabetes, Alzheimers, and cancer.

With this potential to revolutionize medicine, why is it that the president, the pope, governments, and religions worldwide have been tied up for months in the debate over this research? The concern registered by many is this: the most promising stem cells—embryonic stem cells (ES cells)—are harvested from fertilized eggs. Early embryos at the blastocyst stage, four to six days after the first cell division, are destroyed to obtain the stem cells. The fact that current research uses embryos slated for disposal in in vitro fertilization laboratories has not helped untie the ethical knots that have us bound in our current moral impasse.

How should we get through the impasse? It is our considered judgment that not only is this research morally permissible, but that there is an ethical and theological mandate to actively support it. To do otherwise, we have concluded, would be remiss. In short, to not support stem cell research is unethical.

APPEAL TO BENEFICENCE

The principle underwriting our support is beneficence, a bioethical variant of the Christian understanding of agape love. Theological and ethical reflection are at their best when framed by beneficence: an active love of God and neighbor that inspires a celebration of life and a struggle against suffering and death. Beneficence goes beyond nonmaleficence: it goes beyond "do no harm" to look for opportunities to "do good." It defines moral responsibility positively rather than negatively.

The problem with the stem cell debate to date, in our assessment, is that it has been framed by the wrong issue-orienting question. Rather than framing the issue as a version of the abortion issue, which is a negatively framed issue that hides the moral value of a positive agenda, we prefer to frame it as a question regarding the opportunity to improve human health and well-being. The first, though certainly not the only, question we ask of stem cell research is this: Does stem cell research further or hinder the betterment and well-being of humanity? This places the good of those who might benefit from this research as central. The framework of beneficence, rather than the abortion controversy or even nonmaleficence, should serve as the moral guide for this debate.

The argument from beneficence, when applied to the stem cell debate, begins by observing that this form of scientific research promises enormous leaps in the quality of health care. For those who follow Jesus of Nazareth, decisive here is the Nazarene's ministry of healing. Jesus healed. The Christian doctrine of salvation includes healing of body and soul and our relationship with God. We human beings emulate God when we engage in our own ministry of healing. Medical research science, in its own way, contributes to God's healing work on earth. To ignore this divine mandate is itself an ethical concern.

Public support for medical research and health care delivery is a form of agape, selfless love. Sacrificing social and economic resources now is an investment in improving human health in the future; as such it is a form of loving persons we may never meet. Even if those scientists involved are unaware, their's is a divine work.

In short, we uphold that beneficence is the central ethical issue deriving from the public debate over stem cell research.

THE DIGNITY OF THE EMBRYO

Do we concede integrity to our opponents, to those who wish to protect the dignity of the early embryo? Yes we do. Certainly the destruction of embryos for this research is not irrelevant to our ethical considerations. We must ask a question inherited from hundreds of years of ethical reflection, namely, when does life begin? Or better, when does morally relevant personhood begin? Is the fertilized zygote or early embryo an individual human person with dignity?

In *Donum Vitae* in 1987 the Vatican declared that at conception three components make a full human being: sperm, egg, and a divinely implanted soul. However, with advances in embryology such as nuclear transfer (the type of cloning that brought the sheep Dolly into existence), we see that nature does not actually require two parents in the form of sperm and egg. So, these former commonsense pictures of the inception of human life are quickly losing step with scientific understanding of what it takes to make a human individual. Theologians and ethicists must study this rapidly advancing science before ethical conclusions on the status of the embryo are drawn. As the science of embryology increases in complexity, ethical deliberation that relies on outdated knowledge becomes a public embarassment for the church. Simply recasting the stem cell debate as the abortion controversy represents lazy ethics while misleadingly connecting today's scientists with yesterday's pro-choice advocates.

Now, please do not suspect that we are engaging in smoke and mirrors here, trying to hide abortion behind a veiling appeal to complexity. We recognize the seriousness of reasking the question: when does morally relevant personhood begin in embryonic development? Does the fertilized ovum deserve respect; and, if so, what kind of respect? Is the embryo an actual human being? A potential human being? If the embryo must be dismantled to obtain the stem cells within, can we, with the pope stand on the most conservative answer and still support embryonic stem cell research in the name of beneficence? We would say, "yes." Let's reflect for a moment.

SURPLUS EMBRYOS ALREADY DESTINED FOR DISPOSAL

The embryo is a potential human being, to be sure; and respect for the early embryo shows our respect for God's intended future destiny. Beneficence, as we would invoke it here, does not provide blanket support for the economic and political values too often woven into the fabric of scientific research. Because the value of the embryo remains integral to our thinking, we do not support research that would lead to the wholesale fabrication of embryos for research purposes. Opponents of embryonic stem cell research were rightly upset by reports that the Jones Institute for Reproductive Medicine, a fertility clinic in Virginia, had created embryos expressly for medical research.

We support research that uses stem cell lines derived from embryos taken from fertilization labs, embryos that would otherwise be discarded. In the deep freezes of in vitro fertilization clinics we can find vats of liquid nitrogen housing thousands of fertilized zygotes, embryos. Society has already decided to engage in reproductive technology. Excess embryos exist in large numbers. Due to their practical destiny, these surplus embryos will never find connection to a mother's womb, never become a human being.

What should the fate of these embryos be? Should they be kept frozen until freezer burn destroys them? Should they be poured down the drain? Or might a select handful be used in life-giving research? To date, researchers have elected to use only surplus fertilized ova, not to derive new ones slated for destruction, as the pope seems falsely to assume. If and when researchers announce they are creating embryos for destruction, then we'll revisit the pontif's argument.

So, let us ask: Is it ethically licit to take surplus frozen embryos that will never be implanted and press them into the service of life-saving medical research? When the alternative for such already fertilized ova is the drain, is it ethical to sanction this opportunity to provide the human race with what life potential they offer? Armed with the principle of beneficence we want to answer, yes.

BENEFICENCE VERSUS NONMALIFICENCE

We recognize that opponents of embryonic stem cell research make two strong ethical claims. First, they argue from nonmaleficence—that is, "do no harm." Dismantling a blastocyst to obtain stem cells does harm. For opponents, nonmaleficence trumps beneficence. We disagree when destruction of the embryo has already been decided elsewhere. Second, opponents of ES cell research support adult stem cell research. In fact, several organizations opposing embryonic stem cell research have gone so far as to claim that, in light of the burgeoning promise of adult stem cell research, embryonic stem cell research is virtually unnecessary. However, scientifically, this assessment of the relative value of adult stem cells over against embryonic stem cells is simply wrong. Adult stem cells have neither the versatility nor the longevity of ES cells. Research on adult stem cells by no measure serves as a replacement for work on embryonic stem cells. Claims about progress in adult stem cell research, simply put, have been inflated. Certainly commitment to beneficence should include support of adult stem cell research. This support, however, does not diminish the need for continued support of embryonic stem cell research.

Our society finds itself in the middle of moral ambiguity, in an ethical hurricane where nothing seems tied down. Absolutes seem to be coming loose. The task of the ethicist is to render considered judgment, to supply the best recommendation for public policy and scientific protocol to go forward. We hold that stem cell research should be encouraged to go forward so that otherwise discarded early embryos might still provide life-giving potential for our future.

20
Stem Cells and the Catholic Church

Aline H. Kalbian

Strong opposition and condemnation by the Catholic Church is a prominent feature of most contemporary debates about cloning and the use of human embryos in research. So prominent, in fact, that while President Bush was pondering the fate of federal funding for embryonic stem cell research in July of 2001, he met with Pope John Paul II at the Vatican and was urged by him to "reject practices that devalue and violate human life at any stage from conception until natural death."[1] The president's consultation with a religious leader on this issue was certainly remarkable, especially to those concerned with church/state separation. Equally remarkable, however, has been the Church's resolve on this issue.

The particulars of the Church's position on embryonic stem cell research often meld together with a larger set of views opposing abortion, artificial contraception, and most forms of assisted reproductive technology. In fact, in the Pope's comments to President Bush, he explicitly groups several activities under the label of "related evils," asserting that "Experience is already showing how a tragic coarsening of consciences accompanies the assault on innocent human life in the womb, leading to accommodation and acquiescence in the face of other related evils, such as euthanasia, infanticide, and most recently proposals for the creation for research purposes of human embryos, destined to destruction in the process."[2] This slippery slope that begins with abortion leads to evil actions like stem cell research using human embryos. All such actions are connected to what John Paul II calls a "culture of death," and the Church's role in the face of this culture is to proclaim the "gospel of life."

The embryo's status as a human person entitled to full human rights from the moment of conception, including the right not to be harmed, is a centerpiece of this gospel. In this paper, I suggest that while the embryo *is* the central focus of the argument, the image of the embryo represents something much larger; namely a certain understanding of the vulnerable *other*, the neighbor in society. References to the embryo serve as a metaphor that reveals the central Christian message of neighbor love. However, this metaphorical language also hides a set of beliefs about procreation and family. I begin with a brief exploration of how the embryo as metaphor functions. I then set the teaching about the status of the embryo in the broader context of the Catholic Church's view on embryonic stem cell research. Finally, I look at the extent to which the Catholic position is unchanging and absolute. I do this by outlining three elements of Catholic moral reasoning: natural law, the common good, and proportionalism. These three elements all reveal tensions in contemporary Catholic moral thought—tensions that can provide an opening for a different analysis of the use of human embryos in research.

THE EMBRYO AS NEIGHBOR

Pope John Paul II's writings on morality have long echoed the theme of caring for the vulnerable other. In recent decades the image of the embryo has served as the most effective means for capturing the vulnerability of the other who needs protection. On this view, society's acceptance of abortion is intimately related to a larger rejection of "a truly humane future." What becomes clear in much of John Paul II's writing is his view that the Catholic Church ought to speak for those who have no voice or who lack the capacity to speak for themselves.

In *Evangelium Vitae* (EV), the pope's 1995 encyclical on abortion and euthanasia, he describes this care and concern for the other in terms of the "positive requirements" that emerge from the "commandment regarding the inviolability of life."[3] According to EV, the prohibition against killing is a negative commandment; its positive face is the commandment to love the neighbor. As John Paul II reminds his readers, the category neighbor includes not only friends and close relations, but also strangers and enemies. All are neighbors because they are made in the image of God. Thus, the attitude of love *for* the other and the prohibition *against* killing are all part of the same commandment. Interestingly, the attitude of love and reverence for life is intensified from mere attitude to action when the Pope encourages the faithful "to defend and promote life." The ultimate expression of this defense and promotion of life is God's call to humans to "be fruitful and multiply."

The activity of procreation is intimately tied to the protection and defense of life. Just as the neighbor love command is the positive manifestation of the prohibition against killing, procreation is the positive action that correlates with protecting and defending life. Human procreation involves the "special participation of man and woman in the creative work of God."[4] Humans, on this view, cooperate with God who is understood to be present at the moment of conception. This is based on the idea that the new life bears the image and likeness of God. John Paul II writes, "the genealogy of the person is inscribed in the very biology of generation."[5] That genealogy, marked through the *imago dei*, confirms the status of the embryo as person made in God's image. In essence, then, harm to the embryo at any stage is harm to this image and destruction of the embryo destroys the genealogy. Humans as cocreators with God have a special responsibility or mission that requires "accepting and serving life" especially life "when it is at its weakest."[6] This weakness and vulnerability is especially evident in human life at its beginning and its end. Thus, the embryo is in special need of protection.

In the Vatican Instruction, *Donum Vitae* (DV), the Catholic Church asserts the view that while parents do not have a "right" to a child, the child (from the moment of conception) does have a right to "be conceived, carried in the womb, brought into the world, and brought up within marriage."[7] This attribution of rights to the zygote, embryo, and fetus emphasizes the strong connection between marriage and procreation in the tradition. The clear subtext is that the married couple takes on the task of procreation as part of the battle to defend and promote life.

THE CATHOLIC ARGUMENT AGAINST EMBRYONIC STEM CELL RESEARCH

Concern about the treatment of human embryos has long been at the center of Catholic views about reproductive technologies and abortion. The Church has held the unequivocal stance that "Life once conceived, must be protected with the utmost care."[8] Thus, the recent explicit responses to cloning and stem cell research are not new.

The Catholic literature about embryonic stem cell research shares certain salient characteristics. Five are worth noting. The first, and in many ways most significant, is that the embryo from the first moment of fertilization is a human subject with an identity. The most recent Vatican statement on stem cell research identifies this as the primary reason why it is morally illicit to produce and/or use human embryos for the purpose of deriving stem cells. The basis for this attribution of subjectivity and "well-defined identity" is said to be "a complete biological analysis" which leads to the conclusion that from the moment of conception the human embryo "begins its own coordination, continuous and gradual development."[9] This view has been articulated in a broad range of Catholic documents. For example, in the 1987 Vatican Instruction, *Donum Vitae*, we read the following statement: "This teaching remains valid and is further confirmed, if confirmation were needed, by recent findings of human biological science which recognize the zygote (the cell produced when the nuclei of the two gametes have fused) resulting from fertilization the biological identity of a new human individual is already constituted."[10] This quasi-scientific argument is supplemented in Catholic teaching by the belief in the totality or unity of body and spirit. Thus from the very moment of conception, an individuated material entity exists. Since this entity (the newly formed zygote) exists materially as a body, the argument goes, it must also exist as a spirit, and is thus a full human individual.

The notion of an identity is expressed in countless ways through many of the Vatican documents. In the encyclical *Evangelium Vitae*,

Pope John Paul relies on biblical passages for evidence of the subjectivity of the embryo. One of the most often quoted passages, "Before I formed you in the womb I knew you, and before you were born I consecrated you"[11] is mentioned but not explicated in much detail. It is presented as proof that all human beings, even while still in their mother's womb, are "the personal objects of God's loving and fatherly providence"[12] and that "the life of every individual from its very beginning, is part of God's plan."[13] More interesting and elaborate are the references to Job who in the midst of his despair is able to know that "there is a divine plan for his life."[14] Job's trust in God is based partly on his sense of awe at God's creative work. Job's meditation on himself as God's creation is filled with detail. He asks, "Did you not pour me out like milk and curdle me like cheese? You clothed me with skin and flesh, and knit me together with bones and sinews."[15] Of course, this lament is part of an appeal on Job's part for God's mercy. By reminding God that He is the creator who fashioned him, Job is pleading with God. The use of this passage by the Pope as a proof-text that the fetus in the womb is a subject with an identity is interesting because the passage makes no mention of the womb or the moment of conception. It simply affirms God's creative power.

The other passage used by the Pope is the story from the gospel of Luke of the meeting between Elizabeth and Mary who are both pregnant. The significant moment in the story occurs when Mary's greeting to Elizabeth causes Elizabeth's fetus to "leap in her womb." On the Pope's reading, this activity in the womb affirms that "it is precisely the children who reveal the advent of the Messianic age."[16] This evidence of life in the womb is then translated into confirmation of "the indisputable recognition of the value of life from its very beginning."[17] Quoting St. Ambrose of Milan, the Pope highlights the role of "the babies" (John the Baptist and Jesus Christ): "The women speak of grace: The babies make it effective from within to the advantage of their mothers who by a double miracle prophesy under the inspiration of their children."[18] The children or fetuses are exalted and imbued with an almost supernatural power; the mothers are made to seem mere tools in this important transmission. This image is especially effective in the Catholic argument against abortion. It responds to any arguments that characterize abortion as the mother's life vs. the life of the fetus by affirming the view that the fetus takes precedence.

Once the identity of the embryo/fetus is confirmed the next element of the Catholic view follows. It is the philosophical point that this newly formed human subject has rights; namely a right to its own life. In *Evangelium Vitae*, the Pope writes "it must nonetheless be stated that the use of human embryos or fetuses as an object of experimentation constitutes a crime against their dignity as human beings who have a right to the same respect owed to a child once born, just as to every person."[19] While clearly not an argument, this statement emphasizes the sense that at all stages of development, the embryo is a bearer of rights. Thus, from identity as a subject, the move is made to assign full personhood to the embryo. In the language of the Catholic Church the embryo is categorically a human being and a rights-bearing person.

Once the status of the embryo has been determined, both on the basis of biological and philosophical reasoning, the Church then claims that harming the embryo by "irremediably damaging" it or by "curtailing its development is a gravely immoral act and consequently is gravely illicit."[20] This is the moral claim, and it is intricately wound up in the Church's Vatican II teaching about the interdependence of the individual person and the larger community. In *Gaudium et Spes* we see the articulation of this view in the statement, "Man's social nature makes it evident that the progress of the human person and the advance of society itself hinge on each other. For the beginning, the subject and the goal of all social institutions is and must be the human person. . . ."[21] Thus, regardless of the larger benefit that might accrue from stem cell research on embryos, the individual embryo is one and the same as "the human person" for whom social institutions exist.

The fourth element, which is essentially the core of the justification for the above-mentioned moral claim, is stated in the following dictum: A good end cannot justify and evil means. This anticonsequentialist strain in the Catholic tradition serves to rebut the sorts of arguments that weigh the benefits of curing Alzheimer's, Parkinson's, diabetes, and other diseases against the sacrifice of human embryos that have not yet been implanted in a woman's womb. Put differently, the action of harming or destroying an embryo is inherently wrong and its wrongness cannot be lessened in any way by the potential therapeutic procedures that may result.

The fifth and final element of the Catholic argument provides a context by stressing its historical consistency and longevity. This affirmation of history and tradition is an important feature of Catholic moral theology, and is evident in most magisterial documents. The consistency of this teaching serves to remind all that it is not open to revision. This question of the unchanging nature of Catholic doctrine, especially on matters of morality, has recently received quite a bit of attention from Catholic moral theologians. I turn to that in the next section.

CAN CATHOLIC TEACHING ON EMBRYONIC STEM CELL RESEARCH CHANGE?

While this question cannot be responded to easily, it is possible to point to areas of tension in recent debates about Catholic moral theology. Three areas in particular reveal fissures where it might be possible for critics of the tradition to find ways to allow for certain types of embryonic stem cell research. These three areas are (1) critiques of natural law, (2) the rise of social justice arguments, and (3) the legacy of proportionalism.

Natural Law

Natural law theory is practically synonymous with Catholic moral theology, especially with respect to teachings on moral issues related to the body. The belief that human reason's reflections on nature and experience lead to moral knowledge is based on scripture[22] and on the view of the goodness of God's creation. This strong confidence in human reason is evident in the conclusion that natural law enables all persons, not only Catholics, access to moral wisdom and truth. It also affirms the coparticipation of humans in the Divine Law. Human participation in God's plan for creation involves humans in the very act of creation. This emphasis echoes the image of the man and woman's special participation in God's creative work through the event of procreation.

In the decades since Vatican II, a revisionist strain has emerged in Catholicism, and much of its efforts have been directed at revising natural law. This revision has taken a variety of forms. The main problem that many contemporary Catholic theologians who work in medical ethics have noted about traditional natural law is what has been dubbed its physicalist nature. Physicalism refers to the belief that the human body's tendencies are natural and thus from observing these tendencies we can derive moral norms.

Physicalism is best described using the terms—"order of nature" and "order of reason." "Order of nature" refers to the strain of interpretation of the natural law that focuses "on the physical and biological structures given in nature as the source of morality."[23] The other strain of natural law interpretation, "the order of reason" emphasizes natural law as the "human capacity to discover in experience what befits human well-being."[24] Interestingly, while these are identified as two separate strains, many scholars find evidence that Thomas Aquinas accepted both, and as a consequence, claim that contemporary Catholic moral theology ought to incorporate both. This distinction between the two strains of natural law can also be described as the biological vs. the rational strain. The physicalist critique of natural law is directed at an overemphasis or overvaluation of the biological strain. The claim is that deriving moral norms from physical and biological structures and functions violates the totality (unity of body and spirit) of the person. It isolates the corporeality of the person—viewing her just as a body. The critics of an overly physicalist natural law cite the Church's absolute prohibitions on contraception and artificial insemination by husband as two examples of natural law that privileges the biological over the rational.

Pope John Paul II responds to this attack on natural law in contemporary Catholic teaching in his 1993 encyclical letter *Veritatis Splendor*. His response combines the two orders mentioned above into a unified whole which he describes as "man's proper and primordial nature," the "nature of the human person," which is the "*person himself in the unity of soul and body*, in the unity of his spiritual and biological inclinations and of all the other specific characteristics necessary for the pursuit of his end."[25] He wants to retain an emphasis on the moral relevance of the body, but by speaking of the body only in terms of unity and totality, he hopes to avert the accusation of physicalism.

The relevance of this discussion for stem cell research on human embryos is twofold. First, critics of contemporary Catholic teaching can claim that a "physicalist" application of the natural law distorts the full context of embryonic stem cell research. Second, defenders of the tradition can respond using the totality of the body and spirit as evidence that the embryo, even from its earliest stages, is due respect.

Social Justice

The second issue that illuminates this debate is the emphasis in the twentieth century on social justice arguments in Catholic moral theology, especially as they incorporate the notion of the common good. The common good is defined in the Second Vatican Council documents as "the sum of those conditions of social life which allow social groups and their individual members relatively thorough and ready access to their own fulfillment."[26] The sections of *Gaudium et Spes*, the Vatican II *Pastoral Constitution on the Church in the Modern World*, that discuss the common good exhibit the delicate relationship between the individual and society. The Constitution states that "the social order and its development must unceasingly work to the benefit of the human person."[27] The central point is that the social order must be subordinate to the dignity of the individual. This point is central to the Catholic opposition to embryonic stem cell research in the sense that even if the social order could be improved by the use of embryos in research, the assault on the individual's life and dignity (in this case the human embryo) cannot be justified. Nevertheless, it might be possible for some Catholics, especially those who question the full personhood of the embryo, to claim that the common good is benefited by embryonic stem cell research.

Proportionalism

The third interesting element to consider is the debate about proportionalism. Proportionalism is the term used to describe a movement of revisionist theologians who emerged in both the American and European contexts on the heels of the Second Vatican Council in the 1960s. Their primary contention is that the "principle of proportionate reason," which asserts that a proportionality must exist between the means and the end of an action, is an integral part of traditional Catholic moral theology. Essentially, though, the larger project of revision entails a re-evaluation of the object of the moral act. They argue for more attention to the context (both intentions and consequences) of moral actions. Put differently, they claim that the means for achieving an end must be proportionate to that end. Thus, the object of the act, which proportionalists claim has been construed too narrowly by the magisterium, necessarily must include this sense of proportion.

As one might imagine, proportionalism has not been embraced by the magisterium, which has claimed that proportionalism is a form of utilitarianism, despite the fact that proportionalists see themselves as squarely within the tradition of Catholic moral theology. Nevertheless, some have argued that its contributions are more significant than such a dismissal might suggest. Proportionalist approaches would not necessarily lead to full support of cloning and embryonic stem cell research. The call for greater attention to context might, however, provide an opening for Catholics who want to allow for some use of embryonic stem cells and still remain faithful to the tradition.

CONCLUSION

One final issue that perplexes many observers of the Church is its response to research using stem cells from embryos that have already been destroyed. Many find it difficult to place moral blame on the scientist who had nothing to do with the destruction of the embryo and is now using the material for a good end. According to the Church, while destruction and use of embryos are two distinct moral acts, both are morally illicit. In the preceding discussion I focused on the argument directed at the destruction of the embryos. In cases when the embryo has already been produced (for example extra embryos resulting from in vitro fertilization procedures), the moral issue becomes one of cooperation or complicity with evil. The Catholic tradition has a long history of commentary on the question of complicity and cooperation. According to the most recent edition of the Catechism, cooperation in the sin of others includes direct and voluntary participation; praise and approval; failure to disclose or hinder the evil doer; or protection of the evil doer. Thus, any involvement with biological materials from illicitly obtained embryos is also the responsibility of the cooperating individual.

Catholic documents and statements on cloning, stem cell

research, and reproductive technologies all attempt to walk a fine line in their depiction of science and technology. They affirm scientific endeavors that will result in appropriate therapeutic applications, but they want to insure that science and technology are always seen to be in the service of humanity. In the view of the Church, when embryos become tools or objects of scientific study, not only has humanity harmed those particular embryos, but it has also assaulted all that is associated with the vulnerable and weak neighbor.

NOTES

I would like to thank Thomas J. Neal for help in gathering the sources for this paper and Robert R. Cross and John Kelsay for suggestions and comments on an earlier draft.

1. Pope John Paul II, "To the President of the United States of America, H. E. George Walker Bush," [online], www.vatican.va/holy_father/john_paul_ii/speech/2001/july/index.htm [July 23, 2001].
2. Ibid.
3. Pope John Paul II, *Evangelium Vitae* (*The Gospel of Life*) (New York: Random House, 1995), p. 41.
4. Ibid., p. 43.
5. Ibid.
6. Ibid.
7. Congregation for the Doctrine of Faith, "Donum Vitae (Instruction for Human Life in its Origin and the Dignity of Procreation): Replies to Certain Questions of the Day," [online], www.vatican.va/roman_curia/congregations/cfaith/index.htm [February 22, 1987]. See II, A, 2.
8. Ibid., I, 1.
9. Pontifical Academy for Life, "Declaration on the Production and the Scientific and Therapeutic Use of Human Embryonic Stem Cells," [online],www.cin.org/docs/stem- cell-research.html [August 25, 2000], p. 4.
10. Congregation for the Doctrine of Faith, p. 701.
11. Jere. 1:5.
12. Pope John Paul II, *Evangelium Vitae*, p. 61.
13. Ibid., p. 44.
14. Ibid.
15. Job 10:10–11.
16. Pope John Paul II, *Evangelium Vitae*, p. 45.

17. Ibid.

18. Ibid.

19. Ibid., p. 63.

20. Pontifical Academy for Life, p. 4.

21. Pope Paul VI, "*Gaudium et Spes* (Pastoral Constitution on the Church in the Modern World)," [online], www.vatican.va/archive/hist_councils/ii_vatican_council/index.htm [December 7, 1965], p. 25.

22. Rom. 2:14–16.

23. Richard M. Gula, S.S. *Reason Informed by Faith: Foundations of a Catholic Morality* (New York: Paulist Press, 1989), p. 223.

24. Ibid.

25. Pope John Paul II, *Veritatis Splendor* (*The Splendor of Truth*) (Boston: St. Paul Books and Media, 1993), p. 50.

26. Pope Paul VI, "*Gaudium et Spes*," p. 26.

27. Ibid.

21

Testimony of Abdulaziz Sachedina, Ph.D., University of Virginia

Islamic Perspectives in Research with Human Embryonic Stem Cells

Abdulaziz Sachedina for National Bioethics Advisory Commission

Thank you very much for inviting me to give an Islamic perspective on human stem cell research. I do not represent a particular school of thought ("church") in Islam; rather, I speak for the Islamic tradition in general, which is a textual tradition. I have been able to examine a number of primary and secondary sources that have been produced by scholars representing different schools of thought. Two major sects or schools of thought, the Sunni, who form the majority in the Muslim community, and the Shi`ite, who form the minority, do not represent an Orthodox/Reform divide; instead, they are both "orthodox" in the sense that both base their arguments on the same set of texts that are recognized as authoritative by all of their scholars. And yet, it is important to keep in mind the plurality of interpretations displayed by the "traditionalists" and "conservatives" on the one hand, and the "liberals" on the other.

The ethical-religious assessment of research uses of pluripotent stem cells derived from human embryos in Islam can be inferentially deduced from the rulings of the Shari`a, Islamic law, that deal

Reprinted from the National Bioethics Advisory Commission (NBAC).

with fetal viability and the sanctity of the embryo in the classical and modern juristic decisions. The Shari`a treats a second source of cells, those derived from fetal tissue following abortion, as analogically similar to cadaver donation for organ transplantation in order to save other lives, and hence, the use of cells from that source is permissible. For this presentation, I have researched three types of sources in Islamic tradition to assess the legal-moral status of the human embryo: commentaries on the Koranic verses that deal with embryology; works on Muslim traditions that speak about fetal viability; and juridical literature that treats the question of the legal-moral status of the human fetus (*al-janin*).

Historically, the debate about the embryo in Muslim juridical sources has been dominated by issues related to ascertaining the moral-legal status of the fetus. In addition, in order to provide a comprehensive picture representing the four major Sunni schools and one Shi`ite legal school, I have investigated diverse legal decisions made by their major scholars on the status of the human embryo and the related issue of abortion in order to infer religious guidelines for any research that involves the human embryo.

Let me repeat here, as I did when I testified to the Commission about Islamic ethical considerations in human cloning, that since the major breakthrough in scientific research on embryonic stem cells that occurred in November 1998, I have not come across any recent rulings in Islamic bioethics regarding the moral status of the blastocyst from which the stem cells are isolated.

The moral consideration and concern in Islam have been connected, however, with the fetus and its development to a particular point when it attains human personhood with full moral and legal status. Based on theological and ethical considerations derived from the Koranic passages that describe the embryonic journey to personhood developmentally and the rulings that treat ensoulment and personhood as occurring over time almost synonymously, it is correct to suggest that a majority of the Sunni and Shi`ite jurists will have little problem in endorsing ethically regulated research on the stem cells that promises potential therapeutic value, provided that the expected therapeutic benefits are not simply speculative.

The inception of embryonic life is an important moral and social question in the Muslim community. Anyone who has followed Muslim debate over this question notices that its answer has differed at different times and in proportion to the scientific information available to the jurists. Accordingly, each period of Islamic jurisprudence has come up with its ruling (*fatwa*), consistent with the findings of science and technology available at that time. The search for a satisfactory answer regarding when an embryo attains legal rights has continued to this day.

The life of a fetus inside the womb, according to the Koran, goes through several stages, which are described in a detailed and precise manner. In the chapter entitled "The Believers" (24), we read the following verses:

> We created (*khalaqna*) man of an extraction of clay, then We set him, a drop in a safe lodging, then We created of the drop a clot, then We created of the clot a tissue, then We created of the tissue bones, then we covered the bones in flesh; thereafter We produced it as another creature. So blessed be God, the Best of creators (*khaliqin*).[1]

In another place, the Koran specifically speaks about "breathing His own spirit" after God forms human beings:

> Human progeny He creates from a drop of sperm; He fashions his limbs and organs in perfect proportion and breathes into him from His own Spirit (*ruh*). And He gives you ears, eyes, and a heart. These bounties warrant your sincere gratitude, but little do you give thanks.[2]

> And your Lord said to the angels: "I am going to create human from clay. And when I have given him form and breathed into him of My life force (*ruh*), you must all show respect by bowing down before him."[3]

The commentators of the Koran, who were in most cases legal scholars, drew some important conclusions from this and other passages that describe the development of an embryo to a full human person. First, human creation is part of the divine will that determines the embryonic journey developmentally to a human creature. Second, it suggests that moral personhood is a process and achievement at the later stage in biological development of the embryo when God says: "*thereafter* We produced him as another creature." The adverb "thereafter" clarifies the stage at which a fetus attains personhood. Third, it raises questions in Islamic law of inheritance as well as punitive justice, where the rights and indemnity of the fetus are recognized as a person, whether the fetus should be accorded the status of a legal-moral person once it lodges in the uterus in the earlier stage. Fourth, as the subsequent juridical extrapolations bear out, the Koranic embryonic development allows for a possible distinction between a biological and moral person because of its silence over a particular point when the ensoulment occurs.

Earlier rulings on indemnity for homicide in the Shari`a were deduced on the premise that the life of a fetus began with the appreciation of its palpable movements inside the mother's womb, which occurs around the fourth month of pregnancy. In addition to the Koran, the following tradition on creation of human progeny provided the evidence for the concrete divide in pre- and postensoulment periods of pregnancy:

> Each one of you possesses his own formation within his mother's womb, first as a drop of matter for forty days, then as a blood clot for forty days, then as a blob for forty days, and then the angel is sent to breathe life into him.[4]

Ibn Hajar al-`Asqalani (d. 1449) commenting on the above tradition says:

> The first organ that develops in a fetus is the stomach because it needs to feed itself by means of it. Alimentation has precedence over all other functions for in the order of nature growth depends on nutrition. It does not need sensory perception or voluntary movement at this stage because it is like a plant. However, it is given sensation and volition when the soul (nafs) attaches itself to it.[5]

A majority of the Sunni and some Shi`ite scholars make a distinction between two stages in pregnancy divided by the end of the fourth month (120 days) when the ensoulment takes place. On the other hand, a majority of the Shi`ite and some Sunni jurists have exercised caution in making such a distinction because they regard the embryo in the pre-ensoulment stages as alive and its eradication as a sin. That is why Sunni jurists in general allow justifiable abortion within that period, while all schools agree that the sanctity of fetal life must be acknowledged after the fourth month.

The classical formulations based on the Koran and the Tradition provide no universally accepted definition of the term "embryo."

Nor do these two foundational sources define the exact moment when a fetus becomes a moral-legal being. With the progress in the study of anatomy and in embryology, it is confirmed beyond any doubt that life begins inside the womb at the very moment of conception, right after fertilization and the production of a zygote. Consequently, from the earliest stage of its conception, an embryo is said to be a living creature with sanctity whose life must be protected against aggression. This opinion is held by Dr. Hassan Hathout, a physician by training, who was unable to be here today. This scientific information has turned into a legal-ethical dispute among Muslim jurists over the permissibility of abortion during the first trimester and the destruction of unused embryos, which would, according to this information, be regarded as living beings in the in vitro fertilization clinics. Some scholars have called for ignoring the sanctity of fetal life and permitting its termination at that early stage.

A tenable conclusion held by a number of prominent Sunni and Shi`ite scholars suggests that aggression against the human fetus is unlawful. Once it is established that the fetus is alive, the crime against it is regarded as a crime against a fully formed human being. According to these scholars, science and experience have unfolded new horizons that have left no room for doubt in determining signs of life from the moment of conception. Yet, as participants in the act of creating and curing with God, human beings can actively engage in furthering the overall good of humanity by intervening in the works of nature, including the early stages of embryonic development, to improve human health.

The question that still remains to be answered by Muslim jurists in the context of embryonic stem cell research is, When does the union of a sperm and an ovum entail sanctity and rights in the Shari`a? Most of modern Muslim opinions speak of a moment beyond the blastocyst stage when a fetus turns into a human being. Not every living organism in a uterus is entitled to the same degree of sanctity and honor as is a fetus at the turn of the first trimester.

The anatomical description of the fetus as it follows its course from conception to a full human person has been closely compared to the tradition about three periods of forty-day gestation to conclude that the growth of a well-defined form and evidence of voluntary movement mark the ensoulment. This opinion is based on a classical ruling given by a prominent Sunni jurist, Ibn al-Qayyim (d. 1350):

> Does an embryo move voluntarily or have sensation before the ensoulment? It is said that it grows and feeds like a plant. It does not have voluntary movement or alimentation. When ensoulment takes place voluntary movement and alimentation is added to it.[6]

Since there is no unified juridical-religious body representing the entire Muslim community globally, different countries have followed different classical interpretations of fetal viability. Thus, for instance, Saudi Arabia, might choose to follow Ibn Qayyim; while Egypt might follow Ibn Hajar al-`Asqalani. We need to keep in mind that the same plurality of the tradition is operative in North America when it comes to making ethical decisions on any of the controversial matters in medical ethics. Nevertheless, on the basis of all the evidence examined for this testimony, it is possible to propose the following as acceptable to all schools of thought in Islam:

1. The Koran and the Tradition regard perceivable human life as possible at the *later* stages of the biological development of the embryo.
2. The fetus is accorded the status of a legal person only at the later stages of its development, when perceptible form and voluntary movement are demonstrated. Hence, in earlier stages, such as when it lodges itself in the uterus and begins its journey to personhood, the embryo cannot be considered as possessing moral status.
3. The silence of the Koran over a criterion for moral status (i.e., when the ensoulment occurs) of the fetus allows the jurists to make a distinction between a biological and a moral person, placing the latter stage after, at least, the first trimester of pregnancy.

Finally, the Koran takes into account the problem of human arrogance, which takes the form of rejection of God's frequent reminders to humanity that God's immutable laws are dominant in nature and that human beings cannot willfully interfere to cause damage to others. "The will of God" in the Koran has often been interpreted as the processes of nature uninterfered with by human action. Hence, in Islam, research on stem cells made possible by biotechnical intervention in the early stages of life is regarded as an act of faith in the ultimate will of God as the Giver of all life, as long as such an intervention is undertaken with the purpose of improving human health.

NOTES

1. *Koran* 24: 12–14.
2. Ibid. 41: 9–10.
3. Ibid. 38: 72–73.
4. Sahih al-Bukhari and Sahih al-Muslim, *The Book of Destiny* (Qadar).
5. Ibn Hajar al-'Asqalani, *Fath al-Bari fi Sharh al-Sahih al-Bukhari, Kitab al-Qadar*, 11: 482.
6. Ibn al-Qayyim, *Al-Tibyan fi aqsam al-Qur'an* (Cairo, 1933), p. 255.

22

Human Embryonic Stem Cell Research

An Intercultural Perspective

LeRoy Walters

The successful creation of human embryonic stem cells in 1998 opened the door to an important new area of biomedical research .[1] The combination of human embryonic stem cell (HESC) research with the technique of somatic cell nuclear transfer, or cloning—first described in a mammal in 1997[2]—provided scientists with additional research options. In the short run, HESC research—including cloning research—is likely to facilitate the understanding of normal and abnormal cell differentiation and human development, including a better understanding of disease. In the longer term, researchers hope to provide new approaches to therapy for diseases like juvenile onset diabetes or Parkinson's disease and for injuries to the brain or spinal cord.

The novel research techniques reported in 1997 and 1998 gave new impetus to an ethical discussion that had begun in earnest in 1978—the year that the first infant born following in vitro fertilization (IVF) was born. The debate about research involving human embryos

---

From *Kennedy Institute of Ethics Journal* 14, no. 1 (2004): 3–38

was conducted in multiple countries and cultures during the late 1970s, the 1980s, and the early 1990s. HESC research introduced subtle but important changes into this ongoing debate. First, the research involved the cultivation of embryos for a definite number of days and to a distinct stage of preimplantation development—for five days and to the early blastocyst stage. Second, the new stage of human embryo research was not focused primarily on assisting or preventing fertilization but rather on understanding and eventually treating serious diseases and injuries. Third, HESC research provided a new tool for basic science, with broad potential application in genetics and developmental biology. In particular, the technique of introducing human embryonic stem cells into laboratory animals like mice made possible the long-term study of how cells differentiate to become particular types of cell, for example, nerve cells.

Numerous nations, states, cultures, and religious traditions have considered it important to review their policies and moral judgments on human embryo research in the light of these new developments. [This chapter] seeks to review this seven-year discussion, to identify the major policy options that have been adopted, and to examine how major religious traditions have responded to the new technical possibilities while also contributing to the public-policy debate.

Six policy options regarding human embryonic stem cell (HESC) research have been adopted in the various nations and cultures of the world. I begin by summarizing the six options, then attempt to characterize the policies adopted in four major regions of the world: Europe, the Middle East, Asia and the Pacific Rim, and North America. I also review the recent debate at the United Nations on cloning for reproduction and cloning for biomedical research. Finally, I explore how several major religious traditions have responded to the question of HESC research and enquire whether the diverse religious perspectives on HESC research correlate, at least in part, with national policies regarding the research.

SIX POLICY OPTIONS

In the following analysis, I distinguish among six possible policy options regarding human embryo research and HESC research:

Option 1: No human embryo research is permitted, and no explicit permission is given to perform research on existing human embryonic stem cells;

Option 2: Research is permitted only on existing human embryonic stem cell lines, not on human embryos;

Option 3: Research is permitted only on remaining embryos no longer needed for reproduction;

Option 4: Research is permitted both on remaining embryos (see Option 3) and on embryos created specifically for research purposes through in vitro fertilization (IVF);

Option 5: Research is permitted both on remaining embryos (see Option 3) and on embryos created specifically for research purposes through somatic cell nuclear transfer into human eggs or zygotes; and

Option 6: Research is permitted both on remaining embryos (see Option 3) and on embryos created specifically for research purposes through the transfer of human somatic cell nuclei into nonhuman animal eggs, for example, rabbit eggs.

POLICIES BY REGION

Europe

Within Europe, the United Kingdom (U.K.) has, since 1990, adopted the most liberal policies regarding HESC research, accepting both Options 4 and 5. In April 2003, Belgium joined the U.K. and allowed the creation of embryos for research purposes, either through in vitro fertilization or nuclear transfer. Sweden seems to be moving step-by-step toward a similar policy. On the opposite end of the spectrum, several European nations prohibit all human embryo research and do not expressly permit research with already-existing human embryonic stem cells (Option 1). The conservative nations at present include Austria, Ireland, Italy, Norway, and Poland. One nation, Germany, has adopted Option 2; it permits the importation and use of human embryonic stem cells that were derived outside Germany before January 1, 2002 (Option 2, with a time limit). A majority of European nations accept, or are likely to accept, the use for HESC research of remaining (or supernumerary) embryos that are no longer needed for reproduction (Option 3). The Czech Republic, Denmark, Finland, Greece, Hungary, the Netherlands, Russia, and Spain have adopted Option 3, either explicitly or de facto. France and Switzerland are gradually moving toward the acceptance of Option 3—although HESC research is a contentious issue in both of the nations. In Belgium, Denmark, Germany, Greece, Spain, and the U.K., HESC research policies were liberalized between 2001 and 2003. France and Switzerland are likely to adopt more permissive policies in 2004.

Given the division of opinion among European nations, it is not surprising that the European Union (EU) has had difficulty achieving consensus on whether it will *fund* HESC research through its joint 2002–2006 program, officially called the 6th Research Programme. In December 2003, there was a virtual stalemate on this question, with Austria, Germany, Italy, and Portugal favoring a variation on Option 2, and most other EU members advocating Option 3. The European Parliament voted for Option 3 by a 291–235 margin in November 2003, but the ministers of the European Commission have not been able to resolve the issue.[3] The 6th Research Programme of the EU began funding new projects in January 2004, but it is not clear whether HESC research proposals will be eligible for funding under the program. Of course, what research the EU funds or does not fund has no bearing on what research is *legally permitted* within each of the 15 EU member states.

Two additional factors in the European debate are worth noting in passing. Most European nations have considered and some have ratified the April 1997 Council of Europe Convention on Human Rights and Biomedicine. The convention specifies that "the creation of embryos for research purposes is prohibited" (Paragraph 18.2). Thus, nations that have ratified this convention have legally bound themselves not to adopt Option 4, and presumably not to adopt Option 5, as well. Nations like the U.K. that have adopted Options 4 and 5 either do not ratify this convention or formally signal their reservations to provisions like Paragraph 18.2 during the ratification process. After the successful cloning of a sheep in 1997, the Council of Europe adopted, in January 1998, an "Additional Protocol" to this convention "on the Prohibition of Cloning Human Beings." This additional protocol, also ratified by numerous nations in Europe, was limited explicitly to a ban on *reproductive* cloning.

The Middle East

Israel

Since 2000, Israel has been one of the world leaders in HESC research. A law adopted in December 1998 (Law 5759–1999) prohibited reproductive cloning but implicitly permitted research cloning (Option 5). In August 2001, the Bioethics Committee of the National Academy of Sciences and Humanities of Israel recommended that the nation accept Option 5 but reject Option 4, arguing that remaining embryos from infertility clinics should obviate the need to create embryos for research purposes by means of IVF. In response to the Bioethics Committee report, the Ministry of Health empowered the National Helsinki Committee for Genetic Research in Humans to review applications involving either the use of remaining embryos from infertility clinics (Option 3) or the creation of human embryos through somatic cell nuclear transfer (Option 5), provided that the guidelines of the Bioethics Committee report were followed. As of December 2003, the Helsinki Committee had approved one research proposal involving remaining embryos; however, it had not yet approved any protocols involving nuclear transfer. In early 2004, renewal of the December 1998 law, with possible modifications to the duration of the ban on *reproductive* cloning, was being considered by the Knesset.[4]

Iran

During the summer of 2003, Iranian scientists at the Fertility Research Centre, an affiliate of the University Jihad Institute, announced that they had succeeded in creating human embryonic stem cells. This achievement was celebrated by the Iranian government. On September 2, 2003, Ayatollah Ali Seyyed Khamenei, Supreme Leader of the Islamic Revolution, publicly received and congratulated the scientists who had produced the cells.[5]

Asia and the Pacific Rim

In this region, China, for many years, has adopted the most liberal policies. Alone among the world's nations, it has permitted scientists to transfer human nuclei into animal eggs (Option 6). Singapore is moving toward the acceptance of Options 4 and 5, and Japan is debating whether to accept Options 4 and 5, as well as the currently accepted Option 3. Australia, which earlier had been characterized by diverging policies in the various states, achieved a national consensus on Option 3 (with a time limit) in December 2002. India has apparently adopted Option 5, de facto. Taiwan has accepted Option 3 without legislation. . . .

Australia

After several years of debate and in the face of radically differing policies at the state level, the Australian Council of Governments agreed on a unified policy on HESC research in April 2002. The consensus policy, Option 3 with a time limit, was formally enacted into law in two bills adopted in December 2002—after a long and somewhat contentious debate on the precise rules for HESC research. Under the terms of the Research Involving Human Embryos Act 2002, embryos used in research must have been created before April 5, 2002, in order to qualify as "excess" embryos. The law is administered by a nine-member, interdisciplinary Licensing Committee located within the National Health and Medical Research Council. The structure and functions of the Licensing Committee are in most respects analogous to those of the U.K. Human Fertilisation and Embryology Authority.[6]

North America

Canada

Since the completion of a comprehensive report on assisted reproductive technologies by a Royal Commission in November 1993, several sessions of the Canadian Parliament have attempted to enact legislation that would regulate both infertility services and human embryo research. Beginning in 2001, the effort to regulate human embryo research has proceeded along two parallel tracks: the Parliament has sought to pass broad legislation that encompasses various aspects of human reproduction, while the Canadian Institutes of Health Research (CIHR) have focused on the narrower topic of

HESC research. After publishing a discussion paper in March 2001, CIHR published "Human Pluripotent Stem Cell Research: Guidelines for CIHR-Funded Research."[7] The guidelines stipulate that Canadian researchers should use only remaining embryos in HESC research (Option 3). In the fall of 2003, CIHR took the additional step of appointing a Stem Cell Oversight Committee, which has begun evaluating individual research protocols.[8] Meanwhile, a bill entitled "An Act Respecting Assisted Human Reproductive Technologies and Related Research" was approved by the House of Commons in October 2003 and by the Senate in March 2004. The new law will permit the use of remaining embryos in HESC research and will establish the Assisted Human Reproduction Agency of Canada to monitor infertility services and to license and oversee all human embryo research performed in Canada. The Canadian Parliament has thus formally endorsed Option 3.[9]

Mexico

During the 2003 legislative year, the Chamber of Deputies in Mexico's Parliament debated legislation about human cloning. In January 2003, the chair of the Chamber's Health Commission, Maria Eugenia Galvan, announced that she would propose legislation to ban human cloning in Mexico. One month later, deputies from the Democratic Revolutionary Party (PRD) announced their intention to exclude research cloning from any ban on human cloning. The issue was debated by the Deputies in April 2003, and at year's end, on December 3, a bill banning both reproductive and research cloning was approved by the Chamber of Deputies.[10] The President of the Mexican Academy of Sciences and researchers at the National Autonomous University of Mexico (UNAM) responded immediately by vigorously criticizing the Chamber's action.[11]

United States

U.S. policy on HESC research can be described accurately as a patchwork of diverse policies at the state level plus a unified federal policy on the *funding* of such research. At the state level, eleven states have prohibited all human embryo research, including, one assumes, the derivation of human embryonic stem cells through the destruction of human blastocysts (Option 1).[12] Two additional states (Arkansas and Virginia) have banned all human cloning, including research cloning (Option 5 prohibited). However, Option 3 is accepted implicitly in these two states, and Option 4 is not prohibited by their anticloning legislation. Two states (California and New Jersey) explicitly have endorsed research cloning (Option 5) and seem to permit the creation of embryos for research purposes through IVF as well (Option 4).[13] In the remaining thirty-five states, none of the research options is prohibited by law. Thus, it would seem that Option 3, Option 4, Option 5, or even Option 6 would be legally permissible in those thirty-five states.

At the federal level, the current policy was established through a speech given by President George W. Bush on August 9, 2001. President Bush endorsed Option 2, with a time limit on the date by which the human embryonic stem cell lines must have been created—the date of the speech. Initial NIH estimates of sixty to seventy embryonic stem cell lines worldwide proved to be unrealistically high, and at the beginning of 2004 the number of stem cell lines that were well characterized, available to researchers, and derived before August 9, 2001, was officially listed as fifteen.[14]

The United States Congress also has debated the question of human cloning in 2001, 2002, and 2003. Bills that would ban both reproductive and research cloning have been sponsored by Senator Sam Brownback (R-KS) and Congressman Dave Weldon (R-FL) in each of the past three sessions of Congress. This effort to prohibit both types of cloning has been endorsed publicly by the Bush administration. A competing bill that would outlaw reproductive cloning but permit research cloning (Option 5) has been associated most closely with the names of Senators Hatch (R-UT), Kennedy (D-MA), Feinstein (D-CA), and Spector (R-PA). The Weldon bill passed in the House in 2001 and again in 2003 by substantial majorities. However, neither Senate bill seems to enjoy the support of a filibuster-proof majority, and the Senate has not taken action on the issue of cloning.[15] . . .

RELIGIOUS TRADITIONS AND HESC RESEARCH

Three interdisciplinary advisory bodies on HESC research convened in recent years have invited representatives of major religious traditions to comment on the ethics of such research: the U.S. National Bioethics Advisory Commission (NBAC) (1999), the European Group on Ethics of the European Commission (2000), and Singapore's Bioethics Advisory Committee (2001). In the papers and letters submitted by these representatives, a rather consistent picture of the various religious viewpoints emerges. Because human embryo research became an important scientific field only in the late 1960s, the traditions have had less than forty years to wrestle with the questions raised by the research. The time interval since the first publication about the cloning of a mammal (1997) and about the creation of human embryonic stem cells (late 1998) is even more abbreviated. Commentators thus frequently seek to extrapolate from long-debated issues like contraception and abortion when they discuss novel questions like human embryo research, HESC research, and cloning.

Judaism

In the Jewish tradition, moral status is not ascribed to the human embryo at the time of fertilization. Instead, the virtually unanimous view is that the human embryo is "like water" during the first forty days of its development. In the words of Moshe Tendler, an Orthodox scholar,

> The Judeo-biblical tradition does not grant moral status to an embryo before 40 days of gestation. Such an embryo has the same moral status as male and female gametes, and its destruction prior to implantation is of the same import as the "wasting of human seed."[16]

The Jewish religious tradition also places a strong emphasis on the saving of life (*pikuach nefesh*), and several commentators from this tradition have considered the ultimate goal of HESC research to be life saving. Laurie Zoloth, a conservative Jewish scholar, described this constructive mission in her testimony before NBAC.

> The task of healing in Judaism is not only permitted, it is mandated. This [viewpoint] is supported and directed not only in early biblical passages ("you shall not stand idly by the blood of your neighbor," and "you shall surely return what is lost to your neighbor"), but in numerous rabbinic texts as well. The general thrust of Jewish response to medical advances has been positive,

> even optimistic, linked to the notion that advanced scientific inquiry is a part of *tikkun olam*, the mandate to be an active partner in the world's repair and perfection.[17]

The positions argued by Tendler and Zoloth are strongly supportive of Option 3, the derivation of embryonic stem cells from remaining human embryos. Neither scholar explicitly discusses the morality of Options 4 or 5, although Zoloth (J-19) asks whether nuclear transfer would constitute an "improper mixing of two kinds."

Islam

Within Islam one cannot speak of unanimity on the question of embryonic moral status. Various scholars and religious leaders issue their formal opinions, or fatwas, but no single individual or group exercises supreme authority in matters of doctrine or practice.[18] At the same time, however, the overwhelming majority of Muslim legal commentators through the ages have accepted the morality of abortion through either the fortieth day or the fourth month of pregnancy. A classic commentary from the ninth century states the majority position on human embryonic development quite clearly.

> Each one of you possesses his own formation within his mother's womb, first as a drop of matter for forty days, then as a blood clot for forty days, then as a blob for forty days, and then the angel is sent to breathe life into him (*Sahih al-Bukhari* [d. 870] and *Sahih al-Muslim* [d. 875], The Book of Destiny [*qadar*].[19]

This developmental view is quite compatible with the acceptance of HESC research involving five-day-old blastocysts. In a statement before NBAC, Abdulaziz Sachedina summarized the various Islamic faith traditions as follows:

> [O]n the basis of all the evidence examined for this testimony, it is possible to propose the following as acceptable to all schools of thought in Islam:
>
> 1. The Koran and the Tradition regard perceivable human life as possible at the *later* stages of the biological development of the embryo.

2. The fetus is accorded the status of a legal person only at the later stages of its development, when perceptible form and voluntary movement are demonstrated. Hence, in earlier stages, such as when it lodges itself in the uterus and begins its journey to personhood, the embryo cannot be considered as possessing moral status.
3. The silence of the Koran over a criterion for moral status (i.e., when ensoulment occurs) of the fetus allows the jurists to make a distinction between a biological and a moral person, placing the latter stage after, at least, the first trimester of pregnancy.[20]

Sachedina's interpretation of Islam is that the tradition permits the use of five-day-old blastocysts to produce embryonic stem cells (Option 3). It is less clear whether Sachedina considers Options 4 and 5 also to be compatible with Islamic law and ethics.

In Singapore, Muslims constitute almost 15 percent of the population, thus slightly outnumbering Christians. During its deliberations on HESC research, the Singapore Bioethics Advisory Committee solicited the opinion of local Islamic scholars on HESC research. The Legal (*Fatwa*) Committee of the Majlis Ugama Islam Sinapura (Islamic Religious Council) responded as follows:

> The Fatwa Committee rules that the opinion of the Bioethics Advisory Committee to use stem cells from embryos below fourteen days for the purpose of research, which will benefit mankind, is allowed in Islam. This is with the condition that it is not misused for the purpose of human reproductive cloning, which would result in contamination of progeny and the loss of human dignity.[21]

The Fatwa Committee's statement clearly endorses Option 3 and, by implication, accepts research cloning (Option 5), as well.

Buddhism

Buddhists are the largest religious group in Singapore, constituting 42.5 percent of the population. Representatives of this tradition were contacted by the Singapore Bioethics Advisory Committee to solicit their views on HESC research. The Secretary General of the Singapore Buddhist Federation responded on behalf of his tradition, affirming the moral permissibility of the research while expressing reservations about cloning.

> The basic precept of Buddhism is against harming and killing all beings. We are taught to have love and compassion for all beings. Regarding the research on human stem cell[s], Buddhism will look at it seriously from the point of intention. If the intention of the research is to find [cures] specifically to human therapeutic[s]—[i]n other words, if the aim of the research is to help and benefit humankind, then we will deem the research as ethical. On the other hand, if the research is something just for the sake of doing or simply to make money out of it, then we will feel it is unethical.
>
> As for human [cloning], although Buddhism did not state that beings are created by God and . . . different forms of birth are mentioned in the scriptures, . . . we are definitely against it. We feel that this will affect the society both morally and socially.[22]

Whether the Federation's objection to cloning extended to research cloning or focused exclusively on reproductive cloning was not entirely clear in the text of the response. The Federation clearly accepted Option 3 but stipulated that the research must be done with the proper intention. In Buddhist thought worldwide, there is a clear diversity of opinion on HESC research. There is also no central authority to adjudicate ethical disagreements. One of the most vigorous critiques of HESC research was published in 2002 by Damien Keown, Reader in Buddhism at Goldsmiths College in London and co-editor of the *Journal of Buddhist Ethics*. According to Keown,

> Given the emphasis that Buddhism places on the central virtues of knowledge and compassion, the recent advances in scientific understanding and the prospect of the development of cures and treatments which alleviate human suffering are to be welcomed.
>
> At the same time, however, the Buddhist religion places great importance on the principle of *ahimsa*, or non-harming, and therefore has grave reservations about any scientific technique or procedure that involves the destruction of life, whether human or animal. Such actions are prohibited by the First Precept of Buddhism, which prohibits causing death or injury to living creatures.[23]

Keown thus argues that a consistent Buddhist ethic is most compatible with Option 1.[24] In contrast to Keown, Courtney Campbell[25] interprets the principle of *ahimsa* to prohibit only the "infliction of violence or harm on *sentient* beings." Thus, he concludes that the Buddhist tradition in principle could accept research on nonsentient preimplantation embryos under some circumstances (Option 3 or perhaps even Option 5).

Hinduism

Hindus constitute 4 percent of Singapore's population. In response to the Singapore Bioethics Advisory Committee's solicitation in November 2001, the Hindu Endowments Board of Singapore indicated its acceptance of stem cell research within certain time limits. Its position most closely paralleled Option 3. The Board did not specifically address the question of research cloning.

> Energy in the form of life is manifest in the living cells including stem cells derived from early embryos (ES cells).
>
> It is suggested that in Singapore the embryos created by in vitro fertilization, not more than 14 days old, can be used for research.
>
> So also, the ES cells derived from 5 days old frozen embryos can be used to establish the cell lines.
>
> According to our Faith (Hinduism) killing a foetus is a sinful act (BHROONA HATHYA). But whether the 14 day old foetus is endowed with all the qualities of life is not well regarded. Therefore, there is no nonacceptance to use these ES cells to protect human life and advance life by curing disease.[26]

The Board went on to indicate its nonacceptance of embryonic germ cell research because the remainder of the fetus would be killed.[27]

In general, the Hindu ethical tradition has been quite protective of human embryos and fetuses from the time of conception forward. A vivid description of what occurs during sexual intercourse and the initiation of pregnancy appears in the first-century text entitled *Caraka Samhitā*.[28]

> Conception occurs when intercourse takes place in due season between a man of unimpaired semen and a woman whose generative organ, (menstrual) blood and womb are unvitiated—when, in fact, in the event of intercourse thus described, the individual soul (*jiva*) descends into the union of semen and (menstrual) blood in the womb in keeping with the (*karmically* produced) psychic disposition (of the embryonic matter).[29]

Abortion was justified only in the extreme circumstance that the continued development of the fetus threatened the life of the woman.[30] In this tragic circumstance, the quality of mind that was to be exhibited during the destruction of the fetus was *daya*, or compassion.[31]

The Hindu Endowment Board of Singapore explicitly noted that abortion is condemned by the Hindu ethical tradition. The Board seemed to suggest that the destruction of a preimplantation embryo is not equivalent to abortion if the goal of the research being performed is compassionate, that is, directed toward protecting the lives and promoting the health of (other) human beings. In somewhat parallel fashion, Swami Tyagananda, Hindu chaplain at the Massachusetts Institute of Technology (MIT), commented in an April 2002 lecture:

> [T]he question that Hindus may ask is: can the destruction of the embryos in stem cell research be considered as an "extraordinary, unavoidable circumstance" and an act "done for greater good"? If it is, the Hindu tradition will accept the research as ethically justified.[32]

Taoism

Taoists constitute 8.5 percent of Singapore's population. In its submission to the Singapore Bioethics Commission, the Taoist Mission (Singapore) seemed to reject as ethically unacceptable any research that results in the death of living embryos, thus advocating Option 1.

> According to *Laojun jiejing*, "All living creatures that breathe, including those that fly and crawl, should not be killed. Even wriggling creatures also treasure life, even mosquitoes and other insects understand the avoidance of death."[33]
>
> • • •
>
> Taoism treasures life deeply. As indicated by the Taoist saying, "the way of immortality is to value life, and the highest virtue is to save others." Provided that it does not injure life, is not against morality and is not against the teachings of Taoism, Taoism supports research that increases longevity and brings benefit to mankind. Taoism is not supportive of research that goes against the teachings of Taoism, that goes against nature, and that involves the killing of another life, e.g., using embryos for research.[34]

Christianity

Roman Catholicism

The history of Roman Catholic beliefs about the moral status of the developing human being in utero has been studied thoroughly by many scholars. In brief, from Saint Augustine through the nineteenth century (in official church teaching) and the early twentieth century (in canon law), the unformed early fetus was thought to lack a human soul because it lacked sentience. For this reason, contraceptive methods and the termination of a pregnancy before the fortieth day of fetal development were grouped together as sinful but non-homicidal acts. Abortion after the fortieth day—that is, after the ensoulment of the fetus—was considered to be homicide.[35]

Bioethics advisory groups in the U.S., Europe, and Singapore all requested submissions by spokespeople from the Roman Catholic religious tradition. In Europe and Singapore, the Catholic Church spoke with a single voice, reflecting the official teaching of the Vatican. In the U.S., at least one dissenting theological voice was heard.

The European Group on Ethics, in its November 14, 2003, opinion on the ethics of human stem cell research, reproduced the Declaration of the Pontifical Academy for Life dated August 25, 2000. This declaration can be viewed, at least in part, as a formal Roman Catholic response to the U.S. NBAC report of September 1999. Three distinct ethical questions were analyzed: the production of human embryonic stem cells after IVF; the production of human embryonic stem cells through nuclear transfer; and the use of already-existing human embryonic stem cell lines. The declaration emphatically adopts Option 1.

> The *first ethical problem,* which is fundamental, can be formulated thus: *Is it morally licit to produce and/or use living human embryos for the preparation of ES cells?*
>
> *The answer is negative,* for the following reasons:
>
> 1. On the basis of a complete biological analysis, the living human embryo is—from the moment of the union of the gametes—a *human subject* with a well defined identity, which from that point begins its own *coordinated, continuous and gradual development,* such that at no later stage can it be considered as a simple mass of cells.

2. From this it follows that as a "*human individual*" it has the *right* to its own life; and therefore every intervention which is not in favour of the embryo is an act which violates that right. Moral theology has always taught that in the case of "*jus certum tertii*" the system of probabilism does not apply.

3. Therefore, the ablation of the inner cell mass (ICM) of the blastocyst, which critically and irremediably damages the human embryo, curtailing its development, is a *gravely immoral* act and consequently is *gravely illicit*.

4. *No end believed to be good*, such as the use of stem cells for the preparation of other differentiated cells to be used in what look to be promising therapeutic procedures, *can justify an intervention of this kind*. A good end does not make right an action which in itself is wrong.

5. For Catholics, this position is explicitly confirmed by the Magisterium of the Church which, in the Encyclical *Evangelium Vitae*, with reference to the Instruction *Donum Vitae* of the Congregation for the Doctrine of the Faith, affirms: "The Church has always taught and continues to teach that the result of human procreation, from the first moment of its existence, must be guaranteed that unconditional respect which is morally due to the human being in his or her totality and unity in body and spirit: 'The human being is to be respected and treated as a person from the moment of conception; and therefore from that same moment his rights as a person must be recognized, among which in the first place is the inviolable right of every innocent human being to life'" (No. 60).

The *second ethical problem* can be formulated thus: *Is it morally licit to engage in so-called "therapeutic cloning" by producing cloned human embryos and then destroying them in order to produce ES cells?*

The answer is negative, for the following reason: Every type of therapeutic cloning, which implies producing human embryos and then destroying them in order to obtain stem cells, is illicit; for there is present the ethical problem examined above, which can only be answered in the negative.

The *third ethical problem* can be formulated thus: *Is it morally licit to use ES cells, and the differentiated cells obtained from them, which are supplied by other researchers or are commercially obtainable?*

The answer is negative, since: prescinding from the participation—formal or otherwise—in the morally illicit intention of the principal agent, the case in question entails a proximate material cooperation in the production and manipulation of human embryos on the part of those producing or supplying them.[36]

In their testimony before NBAC, three representatives of the Catholic tradition reached differing conclusions. Edmund Pellegrino, a physician, reiterated the official teaching that "human life is a continuum from the one-cell stage until death."[37] However, a Catholic moral theologian, Margaret Farley, and a Catholic moral philosopher, Kevin Wildes, S.J., accented the pluralism of opinion about human embryonic status within the Catholic tradition. Professor Farley expressly dissented from current official Church teaching and argued that the moral case for HESC research is quite powerful, both within the Catholic tradition and in the public forum.

> [A] case *for* human embryo stem cell research can also be made on the basis of positions developed within the Catholic tradition. A growing number of Catholic moral theologians, for example, do not consider the human embryo in its earliest stages (prior to the development of the primitive streak or to implantation) to constitute an individualized human entity with the settled inherent potential to become a human person. The moral status of the embryo is, therefore (in this view), not that of a person, and its use for certain kinds of research can be justified. (Because it is, however, a form of human life, it is due some respect—for example, it should not be bought or sold.) Those who would make this case argue for a return to the centuries-old Catholic position that a certain amount of development is necessary in order for a conceptus to warrant personal status. Embryological studies now show that fertilization ("conception") is itself a process (not a "moment"), and such studies provide support for the opinion that in its earliest stages (including the blastocyst stage, when stem cells would be extracted for purposes of research) the embryo is not sufficiently individualized to bear the moral weight of personhood. Moreover, some of the concerns regarding the use of aborted fetuses as a source for stem cells can be alleviated if safeguards (such as ruling out "direct" donation for this purpose) are put in place—not unlike the restrictions articulated for the general use of fetal tissue for therapeutic transplantation. And finally, concerns about cloning may be at least partially addressed by insisting on an absolute barrier between cloning for research and therapeutic purposes on the one hand and cloning for reproductive purposes on the other (the latter, of course, raising many more serious ethical questions than the former).[38]

Professor Farley accepts Option 3 and seems willing to consider accepting Option 5, as well.

Eastern Orthodoxy

The Eastern Orthodox tradition was represented only in testimony before the U.S. NBAC. Demetrios Demopulos expressed what seems to be a nearly universal consensus within the Orthodox tradition when he argued against the destruction of human embryos for research purposes (Option 1).

> Humans are created in the image and likeness of God and are unique in creation because they are psychosomatic, beings of both body and soul—physical and spiritual. We do not understand this mystery, which is analogous to that of the Theanthropic Christ, who at the same time is both God and a human being. We do know, however, that God intends for us to love Him and grow in relationship to Him and to others until we reach our goal of theosis, or deification, participation in the Divine Life through His grace. We grow in the image of God until we reach the likeness of God. Because we understand the human person as one who is in the image and likeness of God, and because of sin we must strive to attain that likeness, we can say that an authentic human person is one who is deified. Those of us who are still struggling toward theosis are human beings, but potential human persons. We believe that this process toward authentic human personhood begins with the zygote. Whether created in situ or in vitro, a zygote is committed to a developmental course that will, with God's grace, ultimately lead to a human person. The embryo and the adult are both potential human persons, although in different stages of development. As a result, Orthodox Christians affirm the sanctity of human life at all stages of development. Unborn human life is entitled to the same protection and the same opportunity to grow in the image and likeness of God as are those already born.[39]

On the use of existing stem cell lines, Father Demopulos[40] noted that "[w]ishing that something had not been done will not undo it. Established embryonic stem cell lines exist, and their use has great potential benefits for humanity. . . ." He argued that the existing lines should be used "only therapeutically, to restore health and to prevent premature death."[41] Thus, his position closely approximated Option 2.

Protestant Traditions

The bioethics commissions in Europe, Singapore, and the United States all received testimony from representatives of various Protestant traditions. As one might expect, Protestant commentators on the ethics of HESC research did not speak with one voice. In his testimony before the U.S. NBAC, moral theologian Gilbert Meilaender[42] articulated a position that is virtually indistinguishable from that of the Vatican (espousing Option 1), although he did not rely on official statements of the Catholic Church. In contrast, Ronald Cole-Turner[43] agreed with official statements of the United Church of Christ that support human embryo research through the fourteenth day of development—but only in the context of public discussion, public accountability, and a concern for social justice (Option 3, with conditions).

In October 2000, an ecumenical group of Protestant and Orthodox (but not Roman Catholic) thinkers submitted a position paper to the European Group on Ethics. The paper was entitled "Therapeutic Uses of Cloning and Embryonic Stem Cells."[44] After laying out three positions on HESC research, including the "intermediate position" of the Church of Scotland, the ecumenical statement ultimately rejects the use of human embryos as means to ends, even the lofty end of promoting human health.[45] The creation of embryos through nuclear transfer is quite clearly condemned, and the ecumenical group urges that "a priority should be put on nuclear transfer research which aims at avoiding the use of embryos, by direct programming from one adult body tissue to another."[46] This ecumenical statement comes close to adopting Option 1.

In letters to the Singapore Bioethics Advisory Committee, two Protestant groups also expressed ethical objections to the destruction of human embryos in HESC research. The National Council of Churches of Singapore—representing Lutherans, Methodists, Anglicans, and Presbyterians—argued that insofar as "experimentation with embryo[s] . . . necessitates their destruction, . . . it is our considered opinion that the ethical concerns far outweigh the therapeutic potentials."[47] The Singapore Council of Christian Churches, comprised of conservative evangelical Protestants, also opposed the "willful destruction of human embryos for medical research."[48] Thus, both groups advocated Option 1. . . .

CONCLUSION: PROSPECTS FOR THE FUTURE

In the years to come, the local, national, and international debate about HESC research is likely to continue. States and nations in which Option 3 is official policy will continue to host, and in some cases to foster, HESC research that uses remaining embryos. The few countries and states that accept Option 4 and/or Option 5—the U.K., Belgium, China, India, Israel, South Korea, two or more states in the U.S., and (probably) Singapore and Sweden—will continue with these more ethically controversial modes of research. If a major scientific or therapeutic breakthrough occurs with, for example, human research cloning, one can predict that some of the current ethical opposition to this mode of research will diminish. On the other hand, if human somatic cells in fact can be reprogrammed and rendered pluripotent (like the minimally differentiated cells of the blastocyst's inner cell mass), then substantial numbers of researchers may gravitate—perhaps with a sigh of relief—toward research methods that do not require the destruction of early human embryos.

NOTES

I thank the following people who have commented on earlier drafts of this article: Cynthia Cohen, Thomas Eich, Lori Knowles, Alexandre Mauron, and Erik Parens. The following people have provided information for specific sections of the essay: Robert Araujo (U.N., Observer Mission of the Holy See); D. Balasubramanian (India); Zelina Ben-Gershon (Israel); Ole Johan Borge (Norway); Robin Alta Charo (numerous nations); In-Chin Chen (Taiwan); Ole Döring (China); Mostafa Dolatyar (U.N., Mission of the Islamic Republic of Iran); Carlos Fernando Diaz (U.N., Mission of Costa Rica); B.M. Gandhi (India); Line Matthiessen-Guyader (European Commission, Directorate General: Research); Yutaka Hishiyama (Japan); Ahmad Hajihosseini (U.N., Observer Mission of the Organization of the Islamic Conference); Phillan Joung (South Korea); Young-Mo Koo (South Korea); Sylvia Lim (Singapore); Carlos Quesnel Menendez (Mexico); Lisette Ramcharan (Canada); Michel Revel (Israel); Carlos M. Romeo-Casabona (Spain); Christian Steineck (Japan); Linda Tan (Singapore); Adam Thiam (Islamic Fiqh Academy, Saudi Arabia); Carolyn Willson (U.N., Mission of the United States); Laurie Zoloth (Judaism).

1. James A. Thomson et al., "Embryonic Stem Cell Lines Derived from Human Blastocysts," *Science* 282 (1998): 1145–47.

2. I. Wilmut et al. "Viable Offspring Derived from Fetal and Adult Mammalian Cells," *Nature* 385 (1997): 810–13.

3. Raphael Minder, "Parliament Approves EU Stem Cell Research," *Financial Times* (November 20, 2003): 10; "Research: Unable to Reach an Agreement, Ministers Close Debate on Stem Cells," *European Report*, no. 2826. European Information Service (December 6, 2003).

4. Personal communications from Michel Revel, Weizmann Institute of Science, Rehovet, Israel, December 16, 2003, and January 16, 2004, and from Zelina Ben-Gershon, National Helsinki Committee, December 17, 2003, and January 16, 2004. See also Nina Gilbert, "News in Brief," *Jerusalem Post*, January 8, 2004, p. 2; Woo Suk Hwang et al.,"Evidence of a Pluripotent Human Embryonic Stem Cell Line Derived from a Cloned Blastocyst," *Science* (February 12, 2004) [online], http://dbtindia.nic.in/policy/consent.html [December 15, 2003]; Indian Council for Medical Research (ICMR), *Ethical Guidelines for Biomedical Research on Human Subjects* (New Delhi: ICMR, 2000); Japan Economic News Wire, "Government Panel OKs Fertilizing Human Ova for Research," Kyodo News Service, December 12, 2003; Soh Ji-Young, "Bill Passed Banning Human Cloning," *Korea Times*, December 31, 2003, p. 1.

5. British Broadcasting Company, "Iran: Khamene'i Praises Scientists for Progress in Cultivating Stem Cells," BBC Monitoring Middle East–Political, September 2, 2003.

6. Sandra O'Malley, "Fed: Stem Cell Laws Pass through Parliament with Little Fanfare," AAP Newsfeed, AAP Information Services Pty. Ltd., December 11, 2002; Surya Palacios, "Mexico Planea Prohibir Clonacion Terapeutica y Regular Reproduccion Asistida," Agence France Presse–Spanish, December 20, 2003; "Research Involving Human Embryos Act (No. 145, 2002)" [online], http://www.nhmrc.gov.au/embryo/index.htm/ [February 4, 2004].

7. Canadian Institutes of Health Research (CIHR), *Human Pluripotent Stem Cell Research: Guidelines for CIHR-Funded Research* (Ottowa: CIHR, 2002) [online], http://www.cihr-irsc.gc.ca/e/publications/1487.shtml#? [February 4, 2004].

8. Personal communication from Cynthia Cohen, Georgetown University, January 31, 2004; "The Stem Cell Oversight Committee," updated January 28, 2004 [online], http://www.cihr-irsc.gc.ca/e/about/19312.shtml#? [February 4, 2004].

9. Canada Newswire, "Assisted Human Reproductive Legislation Receives Approval of Senate," March 11, 2004.

10. Paso a Senado, "Paso el Senado Iniciativa de Ley Prohibe Clonar Humanos," Spanish Newswire Services, EFE News Services, December 4,

2003; Palacios, "Mexico Planea Prohibir Clonacion Terapeutica y Regular Reproduccion Asistida."

11. Claudia Macedo and Arturo Barba, "Storm over Mexican Cloning Ban," *SciDevNet* (Science and Development Network), December 15, 2003 [online], http://www.scidev.net/News/index.cfm?fuseaction=readnews&itemid=1155&language=1[February 2, 2004].

12. The eleven states, beginning with the northeast, are Maine, Massachusetts, Rhode Island, Pennsylvania, Florida, Louisiana, Michigan, Minnesota, Iowa, North Dakota, and South Dakota. Massachusetts has a "safe harbor" provision that allows human embryo research if it is approved by a local institutional review board and is subsequently submitted to, and not disallowed by, the local district attorney. Iowa prohibits "destructive research" on human embryos.

13. Lori B. Andrews, "Legislators as Lobbyists: Proposed State Regulation of Embryonic Stem Cell Research, Therapeutic Cloning and Reproductive Cloning," in *Monitoring Stem Cell Research: A Report of the President's Council on Bioethics* (prepublication version), appendix E (Washington, D.C.: President's Council on Bioethics, 2004).

14. National Institutes of Health (NIH), "NIH Human Embryonic Stem Cell Registry," 2004 [online], http://stemcells.nih.gov/registry/index.asp/ [February 4, 2004].

15. LeRoy Walters, "Research Cloning, Ethics, and Public Policy" (letter), *Science* 299 (2003): 1661.

16. Moshe David Tendler, "Testimony," in *Religious Perspectives*, vol. 3 of *Ethical Issues in Human Stem Cell Research* (Rockville, Md.: National Bioethics Advisory Committee, 2000), p. H-3.

17. Laurie Zoloth, 2000. "Testimony," in *Religious Perspectives*, pp. J-15–J-16, endnote omitted.

18. Thomas Eich, "Muslim Voices on Cloning," *International Institute for the Study of Islam in the Modern World Newsletter* 12 (June 2003): 38–39; Ethics Committee of the Chinese National Human Genome Center at Shanghai, "Ethical Guidelines for Human Embryonic Stem Cell Research (A Recommended Manuscript)," *Kennedy Institute of Ethics Journal* 14 (2001): 47–54.

19. Abdulaziz Sachedina, "Testimony," in *Religious Perspectives*, p. G-4; Singapore Bioethics Advisory Committee, *Ethical, Legal and Social Issues in Human Stem Cell Research, Reproductive and Therapeutic Cloning* (Singapore: BAC. 2002).

20. Sachedena, "Testimony," pp. G-4–G-5.

21. Singapore Bioethics Advisory Committee, *Ethical, Legal and Social Issues in Human Stem Cell Research, Reproductive and Therapeutic Cloning*, pp. G-3–G-71.

22. Ibid., pp. G-3–G-33.

23. Damien Keown, presentation in an October 2001 symposium entitled "Cellular Division," 2001 [online], http://www.science-spirit.org/webextras/keown.html [February 2, 2004].

24. See also Damien Keown, *Buddhism and Bioethics* (New York: St. Martin's Press, 1995).

25. Courtney Campbell, "Religious Perspectives on Human Cloning," in *Commissioned Papers*, vol. 2 of *Cloning Human Beings* (Rockville, Md: National Bioethics Advisory Committee, 1997), p. D-25.

26. Singapore Bioethics Advisory Committee, *Ethical, Legal and Social Issues in Human Stem Cell Research, Reproductive and Therapeutic Cloning*, p. G-3.

27. Ibid.

28. For a discussion of the dating and importance of the *Caraka Samhitā*, see S. Cromwell Crawford, *Hindu Bioethics for the 21st Century* (Albany: State University of New York Press, 2003), pp. 36–38.

29. Julius J. Lipner, "The Classical Hindu View on Abortion and the Moral Status of the Unborn," in *Hindu Ethics: Purity, Abortion, and Euthanasia*, ed. Harold G. Coward et al. (Albany: State University of New York Press, 1989), pp. 53–54; quoting from Sarirasthana, 3.3.

30. S. Cromwell Crawford, *Dilemmas of Life and Death: Hindu Ethics in a North American Context* (Albany: State University of New York Press, 1995), p. 32.

31. Ibid., p. 32.

32. Swami Tyagananda, "Stem Cell Research: A Hindu Perspective," lecture delivered at MIT, April 24, 2002, p. 2 [online], http://home.earthlink.net/~tyag/ Stem%20Cell%20Research.doc [February 2, 2004].

33. Singapore Bioethics Advisory Committee, *Ethical, Legal and Social Issues in Human Stem Cell Research, Reproductive and Therapeutic Cloning*, pp. G-3–G-8.

34. Ibid., pp. G-3–G-9.

35. John T. Noonan Jr., "An Almost Absolute Value in History," in *The Morality of Abortion: Legal and Historical Perspectives* (Cambridge, Mass.: Harvard University Press, 1970), pp. 1–59.

36. Pontifical Academy for Life, "Production and the Scientific and Therapeutic Use of Human Embryonic Stem Cells," August 25, 2000, reprinted in *Text of the Opinion: Ethical Aspects of Human Stem Cell Research and Uses*, European Group on Ethics in Science and New Technologies [of the European Commission], p. 181; endnote references omitted; emphasis in original.

37. Edmund D. Pellegrino, "Testimony," in *Religious Perspectives*, p. F-3.

38. Margaret A. Farley, "Testimony," in *Religious Perspectives*, p. D-4; endnotes omitted.

39. Demetrios Demopulos, "Testimony," in *Religious Perspectives*, p. B-3; endnotes ommitted.

40. Ibid.

41. Ibid., p. B-4.

42. Gilbert C. Meilaender, "Testimony," in *Religious Perspectives*, pp. E–E-6.

43. Ronald Cole-Turner, "Testimony," in: *Religious Perspectives*, pp. A-1–A-4.

44. Church and Society Commission of the Conference of European Churches, "Therapeutic Uses of Cloning and Embryonic Stem Cells," November 14, 2000, reprinted in *Text of the Opinion: Ethical Aspects of Human Stem Cell Research and Uses*, European Group on Ethics in Science and New Technologies [of the European Commission], pp. 190–98.

45. Ibid., p. 196.

46. Ibid.

47. Singapore Bioethics Advisory Committee, *Ethical, Legal and Social Issues in Human Stem Cell Research, Reproductive and Therapeutic Cloning*, pp. G-3–G-66.

48. Ibid., pp. G-3–G-69.

Part 5
Policy Issues

Editors' Introduction

What are we going to do? Remember, to do nothing is to do something. Frank Young does not want inaction, but he would like a moratorium on human embryo research until some of the ethical issues have been clarified. Kenneth Ryan, to the contrary, thinks we have spent far too much time dithering and doing nothing because special religious interests have blocked funding. The time has now come for action, and that means stem cell research and its application.

Those in favor of research and its application will argue that in a society that already allows a million or more abortions a year for reasons that are often little more than convenience or cleaning up the effects of carelessness, it is hypocritical if naught else to forbid the abortion of fetuses produced especially with the aim of alleviating ill health and disease. Those against abortion will argue that abortion continues to be wrong, no matter what the intentions or consequences. But compromises are not intended to make everyone completely happy, rather to point a way forward. We include two opposed pieces. First, a piece by neurobiologist Maureen Condic who does not think that we now need to support work on embryos with public funds. Second, a piece by philosopher Andrew Siegel that tries to uncover reasons behind a compromise, showing how all but extremists can allow some work on stem cells. "Research that involves the destruction of embryos is permissible where there is good reason to believe that it is necessary to cure life-threatening or severely debilitating diseases."

And we finish the section with a give-and-take between Cindy R. Towns and D. Gareth Jones and Simon Clarke. Towns and Jones try to explain policy relations and Simon Clarke tries to give us ethical consistency with respect to the policy.

QUESTIONS TO THINK ABOUT

(1) How should international policy decisions affect how the United States decides to regulate stem cell research? Or should international decisions be irrelevant to how the United States acts?
(2) Is the time for restraint about stem cell research passed? That is, do we know enough about stem cells and what they can do that we should be able to legislate the kind of things scientists can do with them?
(3) Many consider abortion a woman's policy issue? Is the same true of stem cell research? What are the reasons to think it is? What are the reasons to think that it is more than a woman's issue?
(4) What is the role of government in the regulation of science?
(5) What five policy issues do you think are more important than stem cells issues? Do you think there is too much focus on this issue rather than on other policy issues that affect people? If so, what should a policy of stem cell research be?

23
A Time for Restraint

Frank E. Young

The debate on the use of human embryos for research will be one of the more important issues of the twenty-first century. Unlike recombinant DNA technology, embryonic stem cell research most probably will result in the destruction of living embryos. Many people consider this research immoral, illegal, and unnecessary. Therefore, it is imperative to proceed cautiously. Federal funding of research using human embryos or pluripotent cells derived from them would be inappropriate until further resolution of the ethical issues has been achieved.

The ability to grow human embryonic stem (ES) cells in vitro challenges governments, regulatory agencies, and scientific organizations to define the ethical boundaries of using these cells in research. In the United States, President Clinton charged the National Bioethics Advisory Commission (NBAC) to review the medical and ethical considerations of this technology. In September 1999, the NBAC released the Executive Summary of its report.[1]

While noting the existence of diverse views, it formulated a utilitarian approach, justifying public funding of research with human ES cells on the basis of the potential medical benefits. NBAC's primary concern was whether the "scientific merit and substantial clinical promise of this research justifies federal support, and if so with what restrictions and safeguards." Its ethical concern was focused on restricting the sources of embryos.

After National Institutes of Health director Harold Varmus publicly supported the use of human ES cells for research, based on a decision of the general counsel of the Department of Health and Human Services, seventy members of Congress signed a letter of objection. In a letter to *Science* in March 1999,[2] seventy-three scientists, including sixty-seven Nobel laureates, endorsed Varmus's position, claiming that it protects "the sanctity of life without impeding biomedical research." They noted many promising uses for ES cells, including therapeutic advances and reduction of animal studies and clinical trials needed for drug development. Again, the emphasis was on potential benefits, and the ethics of embryo destruction were not addressed.

The fact that experiments using ES cells, as currently performed, result in killing an embryo cannot be ignored so readily. In the United States, such action is in violation of many state laws that protect the embryo.[3] The disintegration of human embryos or the extraction of cells from blastocysts of human embryos [as has been described in primates][4] for the promised but as yet unrealized benefit of patients disregards concerns about the value of the individual that have already been raised by the prospect of human cloning.[5] The devaluation of humans at the very commencement of life encourages a policy of sacrificing the vulnerable that could ultimately put other humans at risk, such as those with disabilities and the aged, through a new eugenics or euthanasia.

Although there is great scientific interest in ES cell research, other recent advances suggest that adult stem cells may be more widely distributed than heretofore recognized and may thus obviate the need for ES cells.[6] Rather than risking public sanction and mistrust from those concerned with the ethical, legal, and moral status of the embryo, is it not wiser to give more than a passing mention to those concerns and in the meantime to do no harm to living embryos? It may be tempting to pursue a scientific imperative that impels us ever forward, but there are major costs. Regulatory policies and processes should take into account public confidence as well as the classical standards of safety and effectiveness. Our pluralistic society must consider the social, religious, medical, environmental, and scientific interests of its citizens. Once credibility is lost, acceptance is eroded.

Should scientific research be limited only by the value of its potential benefits? And who should make the decisions about the limits? To quote J. A. Robertson,[7] "Society, as the provider of the resources, the bearer of the costs, and the reaper of the benefits, has an overriding interest in the consequences of science, hence an interest in the routes and direction that research takes." Scientists who proclaim First Amendment freedom of inquiry are countered by a public suspicion of an inherent conflict of interest when their research support depends on funding from federal and industrial sources. Therefore, any commission regulating research should be composed of individuals of many persuasions and should include people who have no direct or indirect dependence on public monies.

Observations from the recombinant DNA (rDNA) debate can be useful in considering policies regarding human embryo research. The process of policy development was public, and committees consisted of individuals with diverse expertise, opinions, and backgrounds, including those who opposed rDNA research. Although public debate was contentious, careful analysis of the issues prevailed. The results demonstrated that scientists and the public can successfully work together to decide the appropriate use of public funds and formulate regulatory guidelines.[8]

The following recommendations are made to facilitate a consensus. Every nation conducting ES cell research should develop a national policy. In the United States, a representative commission should be appointed to review ES and adult stem cell research, develop an ethical framework for such research, and communicate with the public. It should also examine the adequacy of current guidelines and regulations for in vitro fertilization. A three-year moratorium on human embryo research should be instituted while the commission completes its work. Sufficient funding for research on human adult stem cells and animal embryonic and germinal stem cells should be provided during the moratorium.[9] International harmonization of guidelines could be accomplished through the Organization for Economic Cooperation and Development.

To rush to approve the destruction of embryos in order to harvest and experiment on ES cells is inadvisable and unnecessary. We should address the ethical concerns first.

NOTES

1. National Bioethics Advisory Commission (NBAC), *Ethical Issues in Human Stem Cell Research, Executive Summary* [online], http://bioethics.gov/cgi-bin/bioeth-counter.pl [September 1999].

2. R. P. Lanza et al., "Science over Politics," *Science* 283 (1999): 1849.

3. C. D. Forsythe, "Human Cloning and the Constitution," *Valparaiso University Law Review* 32 (1998): 491.

4. A. W. S. Chan, "Clonal Propagation of Primate Offspring by Embryo Splitting," *Science* 287 (2000): 317.

5. C. S. Campbell, "Religious Perspectives on Human Cloning," in *Cloning Human Beings*, vol. 2 of *Report and Recommendations of the National Bioethics Advisory Commission* (Rockville, Md.: NBAC, 1997), pp. D3–D60.

6. C. Frisen et al., "Central Nervous System Stem Cells in the Embryo and Adult," *Cellular and Molecular Life Sciences* 54 (1998): 935; C. B. Johansson et al., "Identification of a Neural Stem Cell in the Adult Mammalian Central Nervous System," *Cell* 96 (1999): 25; S. Temple and A. Alvarez-Buylla, "Stem Cells in the Adult Mammalian Central Nervous System," *Current Opinion in Neurobiology* 9 (1999): 135; H. G. Kuhn and C. N. Svendsen, "Origins, Functions, and Potential of Adult Neural Stem Cells," *BioEssays* 21 (1999): 625; M. Luquin et al., "Recovery of Chronic Parkinsonian Monkeys by Autotransplants of Carotid Body Cell Aggregates into Putamen," *Neuron* 22 (1999): 743.

7. J. A. Robertson, "Scientist's Right to Research: A Constitutional Analysis," *Southern California Law Review* 51 (1978): 1278.

8. "Coordinated Framework for Regulation of Biotechnology; Announcement of Policy and Notice for Public Comment," *Federal Register* 51 (1986): 23302.

9. Senate Appropriations Subcommittee on Labor, Health and Human Services, and Education, November 4, 1999, testimony, "Stem Cell Research," three speakers: Strom Thurmond (Senate), Jay Dickey (House of Representatives), and Dr. Frank Young (FDA).

24
The Politics and Ethics of Human Embryo and Stem Cell Research

Kenneth J. Ryan, M.D.

THE PROMISE OF HUMAN EMBRYO AND STEM CELL RESEARCH

Significant advances in cellular and developmental biology have occurred with animal research including the dramatic and successful cloning through adult somatic cell nuclear transfer for the sheep named Dolly in 1997.[1] Such cloning has now occurred in mice and cows[2] as well as sheep, and holds great promise in animal husbandry and for application to the manufacture of genetically designed drugs or proteins in sheep or cow milk for use in human medicine. For the time being, there is little interest and much anxiety in applying cloning to human reproduction and many ethical and safety questions remain to be considered. Much has been accomplished otherwise in human reproduction since 1978, after the first human birth with in vitro fertilization of Louise Brown.[3] With

Reprinted from Kenneth J. Ryan, "The Politics and Ethics of Human Embryo and Stem Cell Research," *Women's Health Issues* 10, no. 3 (May/June 2000): 105–10.

this information as background, the promise of future advances using human embryo research seems bright and includes better diagnosis and treatment for male and female infertility, strategies for preventing congenital defects such as preimplantation genetic diagnosis, as well as studies on early human development, fertilization, and implantation. It is, however, the isolation of human stem cells from the inner cell mass of embryos or primordial germ cells of fetuses and the potential for their differentiation in tissue culture into various other cell types such as nerve or blood or heart cells for use in treatment of disease that has captured the imagination of the public and excited scientists. This also might ultimately decrease dependence on organ transplants, for which there has been a chronic shortage and disputes over regional allocations. The isolation of human stem cells from embryos and fetal primordial germ cells occurred in 1998,[4] and its great promise immediately prompted an ethical discussion about the rules then in place forbidding federal funding for research involving human embryos. Applications for use of this stem cell technique have been proposed for such problems as Parkinson's and Alzheimer's diseases, diabetes, heart disease, spinal cord injury, and metabolic diseases in young children. Furthermore, if the cells for transplanting were derived from the insertion of a patient's somatic cell nucleus into an egg or possibly a pluripotent stem cell, the resulting clonal cell line would be expected to have no tissue incompatibility with the donor of the cell nucleus. This would avoid tissue rejection and the need for chronic use of antirejection drugs. Taking this use of stem cells out of the realm of science fiction, scientists already reported in 1999 on the conversion of stem cells from a mouse embryo into specialized brain cells that corrected a myelin deficiency when injected into mice genetically incapable of synthesizing myelin.[5] Beyond theoretical therapeutic applications, there is also the potential for advances in basic knowledge of normal and cancer cell development that can create new opportunities for application to human medicine. The great promise, broad application, and wide support of scientists for this research are important factors in developing public policy about the role of government in the funding, regulation, and ultimate application of such research despite ethical concerns about the use of human embryos or aborted fetuses as the source for research material.

HISTORY OF REGULATIONS GOVERNING FEDERALLY FUNDED RESEARCH ON FETUSES AND EMBRYOS

After the *Roe* v. *Wade* supreme court decision in 1973, there was increased concern in the U.S. Congress that live fetuses in pregnancies destined for abortion would be exploited for research purposes and that abortions might be encouraged largely to satisfy research needs. Before this, there had been reports about abusive research funded by the National Institutes of Health on midtrimester live fetuses both in this country and overseas. As a consequence, a moratorium on federal funding of fetal research was imposed until a national commission created by the National Research Act of 1974 could consider how such abuses could be prevented. The commission report in 1975 provided strict regulatory guidelines for fetal research and the moratorium was lifted.[6] At this time, support for research on tissue derived from deceased aborted fetuses was acceptable. No decision was made on the then prohibited federal funding of research on in vitro fertilization and preimplantation embryos because this was to be reviewed by an ethics advisory board to be appointed by the Secretary of Health, Education, and Welfare. The Ethics Advisory Board was formed two years later in 1977 and issued a report in 1978 after the first in vitro fertilization baby was born in England. The Ethics Advisory Board concluded that research on very early embryos within the first fourteen days of development was acceptable to develop techniques for in vitro fertilization if the gametes were derived from a married couple.[7] (The requirement in 1978 for a couple to be married to provide gametes for pregnancy seems a quaint touch in the present era.) Funding for such research on the early embryo was blocked by pro-life members of Congress and the administrations of both Presidents Ronald Reagan and George Bush, who objected to the inevitable loss of early embryos in the process of developing safe techniques. The favorable report was never implemented and the Ethics Advisory Board was disbanded, resulting in a de facto moratorium on embryo research until a Democratic Congress and president were elected in 1992.

In 1987 and 1988, research on the use of fetal tissue from deceased aborted fetuses for transplantation into patients suffering from Parkinson's disease became an issue and a moratorium was imposed on federal funding until an advisory panel to the director of the National Institutes of Health (NIH) could consider the matter. The chief ethical concerns were whether this would encourage abortion and whether use of such tissue derived from aborted fetuses was unethical because of possible complicity with the abortion. The advisory panel voted nineteen-to-two that with safeguards, federal funding for research on fetal use was appropriate.[8] Nevertheless, proposed legislation to permit such funding was vetoed by President Bush and funding for the use of fetal tissue did not resume until a new Congress and administration were elected and changed the regulations in 1993. Also in 1993, the National Institutes of Health Revitalization Act rescinded the requirement for an ethics advisory board to review research involving preimplantation embryos. A Human Embryo Research Panel was appointed by the director of the NIH to consider whether research on embryos should be conducted, and if so, what review and guidelines were needed for the research to proceed. There was a request for three categories: research acceptable for federal funding, research needing more review before a decision could be made, and research unacceptable for federal funding. The panel recommendation allowed, for example, the use of spare embryos not needed after an *in vit*ro fertilization cycle when consent from gamete donors had been obtained. The permissible research categories included the development of embryonic stem cells as well as work on fertilization, improving pregnancy outcomes, and freezing of oocytes. A type of research not considered acceptable was human cloning for purposes of reproduction. Needing more review, for example, was research on stem cell development from embryos created for that purpose.[9] President Clinton quickly determined that embryos should not be created specifically for research purposes and immediately prohibited funding for this category of research. In the end, funding of any research found acceptable by the panel has not been possible because Congress has regularly added a rider to the annual budget of the NIH prohibiting any research that results in the destruction of an embryo. A strong stand on permitting embryo research to proceed by a broad consensus of the scientific community did not occur until the promise of stem cell research became appreciated. The NIH is currently considering getting around the congressional ban on embryo research by funding research on human stem cells after they are formed, but not funding the actual isolation of the stem cells from embryos which would be accomplished in the private sector.[10] Congressmen are not pleased with this political maneuver, which they consider dealing with the letter and not the intent of their legislation. It is interesting that objections to embryo research could not be overcome for study of problems related to women's reproductive interests, but required the more general promise of the exciting use of stem cells to forge a political will to fight for the research.

In fact, there has been a moratorium on federal funding of embryo research for twenty years, and consequently all improvements in assisted reproduction have to be derived from research funded by the private sector. This has created a sort of moral vacuum with no federal oversight and few research guidelines. During this period, women's interests in advances in reproductive science have been hostage to a conservative agenda in government. Even dues owed by the United States to the United Nations have been withheld for years by a conservative pro-choice Congress. It will be interesting to see who prevails in the standoff over stem cell research: the NIH backed by a broad science consensus or a coalition of seventy pro-life congressmen. Federal support for research involving aborted fetuses finally was regained when the promise of tissue transplantation for Parkinson's and other diseases became great enough to marshal a political will for change.

ETHICAL CONSIDERATIONS OF HUMAN EMBRYO AND STEM CELL RESEARCH

Public discourse on the ethics of abortion started in earnest and became mean-spirited after the Supreme Court 1973 *Roe* v. *Wade* decision. What had been a shadowy private and often illegal matter came into full public view and entered the political mainstream. The issue of research on aborted fetuses became a prominent part of the congressional concern in 1973 lest it increase the number of abortions or contribute to a sense of good derived from an immoral act.

The ethical arguments are by now familiar. One formula is that life is sacred and begins at conception. Therefore, abortion is always wrong unless perhaps the pregnant woman's life is in danger, although some pro-life activists deny an exemption even for threats to the woman's life. The counterargument states that the need for respect of fetal life is not absolute. At fertilization the beginning life starts with a single cell without a nervous system, feelings, interests, or rights, and the pregnant woman's needs and wants should be controlling. During the first trimester, the primitive status of the fetus progressively changes. As the fetus develops, its relative standing should change, and after viability it starts assuming status in society until at birth it assumes full protection under the law. Between these two extremes are many formulations including the familiar and more broadly accepted justifications for abortion. Examples are that the pregnancy threatens a woman's life or health, including mental health, and reasons including rape, incest, and severe fetal defect. The right to abortion in the first trimester seems widely accepted. In addition, the conventional wisdom is that early abortions are preferable on all counts, but after viability the justification for an abortion should be more stringent. All of those arguments have been challenged on the basis of whether they are religious in nature and therefore are not controlling in a pluralistic secular society. In addition, all sorts of metaphors and rhetoric have been employed to sway public opinion, such as whether the right and wrong of abortion are similar to slavery in America before Lincoln emancipated the slaves, or like the Nazi holocaust; this method of persuasion further inflames the passions in the dispute.[11] Although the rhetoric may seem superficially appealing, most of the analogies simply to do not stand up to careful scrutiny. It has also been denied that there are many true cases of rape or incest and that women lie about this to obtain an abortion. This attack on the rape victim's claim seems part of a cruel and pervasive history of sex discrimination in a male-dominated world in the face of the public record on abuse of women. The idea that women ought to be responsible for their sexual acts and should pay the price if they get pregnant is also part of the old repressive puritanical response for what is presumed immoral behavior.[12]

The use of an abortion argument to prevent research on the dead fetus or the use of tissue for treating problems such as Parkinson's disease is part of an agenda to keep abortion a political issue. The use of tissue for these purposes from spontaneously aborted fetuses, or from deceased children and adults, although less often useful, has not engendered the same fierce argument. On the debate over whether one can use fetal tissue or conduct research after an induced abortion, the argument has been that the use of tissue from aborted fetuses will increase the numbers of abortion or justify them, and that in any case the researchers become complicit in an immoral act. Finally, it has been claimed that a woman seeking an abortion should not have the right to give consent to the disposition of the fetus; the counterargument is that although the rationale of this claim is denied, it is not clear who else has standing to give or withhold consent.

Arguments about the ethics of research on the preimplantation embryo and the work on stem cells derived from embryos or fetuses follow those used against abortion, except the conflict between a pregnant woman's rights and status of the fetus she carries are not relevant. The argument here is about the moral status of the developing embryo in the laboratory, usually in the first week or two of development, where the issue of sacredness and rights and interests are raised. Again at issue is the question of the absolute sanctity of all life even at its inception versus a symbolic regard for life that grows with its development. Some, including President Clinton, could not justify creating an embryo in vitro just for research; hence, his directive forbidding federal support for this. However, most people believe that embryos created for infertility that are no longer needed and otherwise would be discarded could be used for research to advance knowledge in reproduction and benefit humankind generally. The offshoot of federal support for stem cell development from embryos seems a reasonable extension of this belief. The complicity argument of automatic responsibility for the loss of an embryo destined for discard anyway seems as far-fetched as complicity for abortion of a fetus with the subsequent use of its tissue when the events are kept separate by law and regulation.[13] It should be obvious that assisted reproduction technologies may produce excess embryos in infertility cycles which can be frozen for later use, but discard of such excess embryos ultimately will be a problem whether or not they are used for research.

None of the ethical arguments seem destined to change made-up minds on the fundamentals of how sacred and absolute life in the womb or test tube is at conception and immediately thereafter. Claims on behalf of the embryo compete with a woman's right to control her reproductive interests, and with society's pursuit of beneficial research in the areas of reproduction and general biology and medicine. Arguments based on principle on the status of life at its beginning have not changed many minds and are unlikely to in the future. There is only an appeal to a point of view and resulting convictions. In such a case, common sense and the consequences of a course of action should be controlling. We never have and are unlikely ever to grant full citizenship and personhood to the fertilized egg or contingent life before viability. The benefits from carefully regulated research are undeniably enormous. Arguments that this is a cause of abortions or a complicity in immoral acts are not convincing. In my view, public policy should favor governmental support for embryo and stem cell research.

NOTES

1. I. Wilmut et al., "Viable Offspring Derived from Fetal and Adult Mammalian Cells," *Nature* 385 (1998): 810–13.

2. T. Wakayama et al., "Full-Term Development of Mice from Enucleated Oocytes Injected with Cumulus Cell Nuclei," *Nature* 394 (1998): 369–74; Y. Kato et al., "Eight Calves Cloned from Somatic Cells of a Single Adult," *Science* 282 (1998): 2095–98.

3. P. C. Steptoe and R. G. Edwards, "Birth after Reimplantation of a Human Embryo," *Lancet* 2 (1978): 366.

4. J. A. Thomson et al., "Embryonic Stem Cell Lines Derived from Human Blastocysts," *Science* 282 (1998): 1145–47; M. J. Shamblott et al., "Derivation of Pluripotent Stem Cells from Cultured Human Primordial Germ Cells," *Proceedings of the National Academy of Sciences USA* 95 (1998): 13726–31.

5. N. Wade, "Mouse Cells Are Converted into Brain Cells," *New York Times*, July 30, 1999, A13.

6. National Commission for the Protection of Human Subjects of Biomedical and Behavioral Research, *Report and Recommendations: Research on the Fetus* (Washington, D.C.: Department of Health Education and Welfare, 1975).

7. J. C. Fletcher and K. J. Ryan, "Federal Regulations for Fetal Research: A Case for Reform," in *Fetal Diagnosis and Therapy*, ed. M. I. Evans et al. (Philadelphia: J. B. Lippincott, 1989).

8. National Institutes of Health (NIH), *Report of the Panel on Fetal Tissue* Transplantation Research (Bethesda, Md.: NIH, 1989).

9. NIH Human Embryo Research Panel, *Report of the Human Embryo Research Panel*, 2 vols. (Bethesda, Md.: NIH, 1994).

10. N. Wade, "Advisory Panel Votes for Use of Embryonic Cells in Research," *New York Times*, June 29, 1999, A13.

11. C. M. Condit, *Decoding Abortion Rhetoric* (Urbana: University of Illinois Press, 1990).

12. K. J. Ryan, "Abortion or Motherhood, Madness and Suicide," *American Journal of Obstetrics and Gynecology* 166 (1992): 1029–36.

13. J. A. Robertson, "Ethics and Policy in Embryonic Stem Cell Research," *Kennedy Institute of Ethics Journal* 9 (1999): 109–36.

25
The Basics about Stem Cells

Maureen L. Condic

In August of last year, President Bush approved the use of federal funds to support research on a limited number of existing human embryonic stem cell lines. The decision met with notably mixed reactions. Proponents of embryonic stem cell research argue that restricting federal funding to a limited number of cell lines will hamper the progress of science, while those opposed insist that any use of cells derived from human embryos constitutes a significant breach of moral principles. It is clear that pressure to expand the limits established by the president will continue. It is equally clear that the ethical positions of those opposed to this research are unlikely to change.

Regrettably, much of the debate on this issue has taken place on emotional grounds, pitting the hope of curing heartrending medical conditions against the deeply held moral convictions of many Americans. Such arguments frequently ignore or mischaracterize the scientific facts. To arrive at an informed opinion on human

---

Reprinted from Maureen L. Condic, "The Basics about Stem Cells," *First Things* (January 2002): 30–34.

embryonic stem cell research, it is important to have a clear understanding of precisely what embryonic stem cells are, whether embryonic stem cells are likely to be useful for medical treatments, and whether there are viable alternatives to the use of embryonic stem cells in scientific research.

Embryonic development is one of the most fascinating of all biological processes. A newly fertilized egg faces the daunting challenge of not only generating all of the tissues of the mature animal but organizing them into a functionally integrated whole. Generating a wide range of adult cell types is not an ability unique to embryos. Certain types of tumors called teratomas are extraordinarily adept at generating adult tissues, but unlike embryos, they do so without the benefit of an organizing principle or blueprint. Such tumors rapidly produce skin, bone, muscle, and even hair and teeth, all massed together in a chaotic lump of tissue. Many of the signals required to induce formation of specialized adult cells must be present in these tumors, but unlike embryos, tumors generate adult cell types in a hopelessly undirected manner.

If a developing embryo is not to end up a mass of disorganized tissues, it must do more than generate adult cell types. Embryos must orchestrate and choreograph an elaborate stage production that gives rise to a functional organism. They must direct intricate cell movements that bring together populations of cells only to separate them again, mold and shape organs through the birth of some cells and the death of others, and build ever more elaborate interacting systems while destroying others that serve only transient, embryonic functions. Throughout the ceaseless building, moving, and remodeling of embryonic development, new cells with unique characteristics are constantly being generated and integrated into the overall structure of the developing embryo. Science has only the most rudimentary understanding of the nature of the blueprint that orders embryonic development. Yet, recent research has begun to illuminate both how specific adult cells are made as well as the central role of stem cells in this process.

The term "stem cell" is a general one for any cell that has the ability to divide, generating two progeny (or "daughter cells"), one of which is destined to become something new and one of which replaces the original stem cell. In this sense, the term "stem" identifies these cells as the source or origin of other, more specialized cells.

There are many stem cell populations in the body at different stages of development. For example, all of the cells of the brain arise from a neural stem cell population in which each cell produces one brain cell and another copy of itself every time it divides. The very earliest stem cells, the immediate descendants of the fertilized egg, are termed embryonic stem cells, to distinguish them from populations that arise later and can be found in specific tissues (such as neural stem cells). These early embryonic stem cells give rise to all the tissues in the body, and are therefore considered "totipotent" or capable of generating all things.

While the existence of early embryonic stem cells has been appreciated for some time, the potential medical applications of these cells have only recently become apparent. More than a dozen years ago, scientists discovered that if the normal connections between the early cellular progeny of the fertilized egg were disrupted, the cells would fall apart into a single cell suspension that could be maintained in culture. These dissociated cells (or embryonic stem cell "lines") continue to divide indefinitely in culture. A single stem cell line can produce enormous numbers of cells very rapidly. For example, one small flask of cells that is maximally expanded will generate a quantity of stem cells roughly equivalent in weight to the entire human population of the earth in less than sixty days. Yet despite their rapid proliferation, embryonic stem cells in culture lose the coordinated activity that distinguishes embryonic development from the growth of a teratoma. In fact, these early embryonic cells in culture initially appeared to be quite unremarkable: a pool of identical, relatively uninteresting cells.

First impressions, however, can be deceiving. It was rapidly discovered that dissociated early embryonic cells retain the ability to generate an astounding number of mature cell types in culture if they are provided with appropriate molecular signals. Discovering the signals that induce the formation of specific cell types has been an arduous task that is still ongoing. Determining the precise nature of the cells generated from embryonic stem cells has turned out to be a matter of considerable debate. It is not at all clear, for example, whether a cell that expresses some of the characteristics of a normal brain cell in culture is indeed "normal"—that is, if it is fully functional and capable of integrating into the architecture of the brain without exhibiting any undesirable properties (such as malignant growth).

Nonetheless, tremendous excitement accompanied the discovery of dissociated cells' generative power, because it was widely believed that cultured embryonic stem cells would retain their totipotency and could therefore be induced to generate all of the mature cell types in the body. The totipotency of cultured embryonic stem cells has not been demonstrated and would, in fact, be difficult to prove. Nonetheless, because it is reasonable to assume embryonic stem cells in culture retain the totipotency they exhibit in embryos, this belief is held by many as an article of faith until proven otherwise.

Much of the debate surrounding embryonic stem cells has centered on the ethical and moral questions raised by the use of human embryos in medical research. In contrast to the widely divergent public opinions regarding this research, it is largely assumed that from the perspective of science there is little or no debate on the matter. The scientific merit of stem cell research is most commonly characterized as "indisputable" and the support of the scientific community as "unanimous." Nothing could be further from the truth. While the scientific advantages and potential medical application of embryonic stem cells have received considerable attention in the public media, the equally compelling scientific and medical disadvantages of transplanting embryonic stem cells or their derivatives into patients have been ignored.

There are at least three compelling scientific arguments against the use of embryonic stem cells as a treatment for disease and injury. First and foremost, there are profound immunological issues associated with putting cells derived from one human being into the body of another. The same compromises and complications associated with organ transplant hold true for embryonic stem cells. The rejection of transplanted cells and tissues can be slowed to some extent by a good "match" of the donor to the patient, but except in cases of identical twins (a perfect match), transplanted cells will eventually be targeted by the immune system for destruction. Stem cell transplants, like organ transplants, would not buy you a "cure"; they would merely buy you time. In most cases, this time can only be purchased at the dire price of permanently suppressing the immune system.

The proposed solutions to the problem of immune rejection are either scientifically dubious, socially unacceptable, or both. Scientists have proposed large scale genetic engineering of embryonic stem cells to alter their immune characteristics and provide a better match for the patient. Such a manipulation would not be trivial; there is no current evidence that it can be accomplished at all, much less as a safe and routine procedure for every patient. The risk that genetic mutations would be introduced into embryonic stem cells by genetic engineering is quite real, and such mutations would be difficult to detect prior to transplant.

Alternatively, the use of "therapeutic cloning" has been proposed. In this scenario, the genetic information of the original stem cell would be replaced with that of the patient, producing an embryonic copy or "clone" of the patient. This human clone would then be grown as a source of stem cells for transplant. The best scientific information to date from animal cloning experiments indicates that such "therapeutic" clones are highly likely to be abnormal and would not give rise to healthy replacement tissue.

The final proposed resolution has been to generate a large bank of embryos for use in transplants. This would almost certainly involve the creation of human embryos with specific immune characteristics ("Wanted: sperm donor with AB+ blood type") to fill in the "holes" in our collection. Intentionally producing large numbers of human embryos solely for scientific and medical use is not an option most people would be willing to accept. The three proposed solutions to the immune problem are thus no solution at all.

The second scientific argument against the use of embryonic stem cells is based on what we know about embryology. In an opinion piece published in the *New York Times*[1] a noted stem cell researcher, Dr. David Anderson, relates how a seemingly insignificant change in "a boring compound" that allows cells to stick to the petri dish proved to be critical for inducing stem cells to differentiate as neurons. There is good scientific reason to believe the experience Dr. Anderson describes is likely to be the norm rather than a frustrating exception. Many of the factors required for the correct differentiation of embryonic cells are not chemicals that can be readily "thrown into the bubbling cauldron of our petri dishes." Instead, they are structural or mechanical elements uniquely associated with the complex environment of the embryo.

Cells frequently require factors such as mechanical tension, large scale electric fields, or complex structural environments provided by their embryonic neighbors in order to activate appropriate genes and maintain normal gene-expression patterns. Fully reproducing these nonmolecular components of the embryonic environment in a petri dish is not within the current capability of experimental science, nor is it likely to be so in the near future. It is quite possible that even with "patience, dedication, and financing to support the work," we will never be able to replicate in a culture dish the nonmolecular factors required to get embryonic stem cells "to do what we want them to."

Failing to replicate the full range of normal developmental signals is likely to have disastrous consequences. Providing some but not all of the factors required for embryonic stem cell differentiation could readily generate cells that appear to be normal (based on the limited knowledge scientists have of what constitutes a "normal cell type") but are in fact quite abnormal. Transplanting incompletely differentiated cells runs the serious risk of introducing cells with abnormal properties into patients. This is of particular concern in light of the enormous tumor-forming potential of embryonic stem cells. If only one out of a million transplanted cells somehow failed to receive the correct signals for differentiation, patients could be given a small number of fully undifferentiated embryonic stem cells as part of a therapeutic treatment. Even in very small numbers, embryonic stem cells produce teratomas, rapidly growing and frequently lethal tumors. (Indeed, formation of such tumors in animals is one of the scientific assays for the "nultipotency" of embryonic stem cells.) No currently available level of quality control would be sufficient to guarantee that we could prevent this very real and horrific possibility.

The final argument against using human embryonic stem cells for research is based on sound scientific practice: We simply do not have sufficient evidence from animal studies to warrant a move to human experimentation. While there is considerable debate over the moral and legal status of early human embryos, this debate in no way constitutes a justification to step outside the normative practice of science and medicine that requires convincing and reproducible evidence from animal models prior to initiating experiments on (or, in this case, with) human beings. While the "potential promise" of embryonic stem cell research has been widely touted, the data supporting that promise is largely nonexistent.

To date there is *no* evidence that cells generated from embryonic stem cells can be safely transplanted back into adult animals to restore the function of damaged or diseased adult tissues. The level of scientific rigor that is normally applied (indeed, legally required) in the development of potential medical treatments would have to be entirely ignored for experiments with human embryos to proceed. As our largely disappointing experience with gene therapy should remind us, many highly vaunted scientific techniques frequently fail to yield the promised results. Arbitrarily waiving the requirement for scientific evidence out of a naive faith in "promise" is neither good science nor a good use of public funds.

Despite the serious limitations to the potential usefulness of embryonic stem cells, the argument in favor of this research would be considerably stronger if there were no viable alternatives. This, however, is decidedly not the case. In the last few years, tremendous progress has been made in the field of adult stem cell research. Adult stem cells can be recovered by tissue biopsy from patients, grown in culture, and induced to differentiate into a wide range of mature cell types.

The scientific, ethical, and political advantages of using adult stem cells instead of embryonic ones are significant. Deriving cells from an adult patient's own tissues entirely circumvents the problem of immune rejection. Adult stem cells do not form teratomas. Therapeutic use of adult stem cells raises very few ethical issues and completely obviates the highly polarized and acrimonious political debate associated with the use of human embryos. The concern that cells derived from diseased patients may themselves be abnormal is largely unwarranted. Most human illnesses are caused by injury or by foreign agents (toxins, bacteria, viruses, etc.) that, if left untreated, would affect adult and embryonic stem cells equally. Even in the minority of cases where human illness is caused by genetic factors, the vast majority of such illnesses occur relatively late in the patient's life. The late onset of genetic diseases suggests such disorders would take years or even decades to re-emerge in newly generated replacement cells.

In light of the compelling advantages of adult stem cells, what is the argument against their use? The first concern is a practical one: Adult stem cells are more difficult than embryonic ones to grow in culture and may not be able to produce the very large numbers of cells required to treat large numbers of patients. This is a relatively trivial objection for at least two reasons. First, improving the proliferation rate of cells in culture is a technical problem that science is quite likely to solve in the future. Indeed, substantial progress has already been made toward increasing the rate of adult stem cell proliferation. Second, treating an individual patient using cells derived from his own tissue ("autologous transplant") would not require the large numbers of cells needed to treat large populations of patients. A slower rate of cell proliferation is unlikely to prevent adult stem cells from generating sufficient replacement tissue for the treatment of a single patient.

The more serious concern is that scientists don't yet know how many mature cell types can be generated from a single adult stem cell population. Dr. Anderson notes, "Some experiments suggest these [adult] stem cells have the potential to make midcareer switches, given the right environment, but in most cases this is far from conclusive." This bothersome limitation is not unique to adult stem cells. Dr. Anderson goes on to illustrate that in most cases the evidence suggesting scientists can induce embryonic stem cells to follow a specific career path is equally far from conclusive. In theory, embryonic stem cells appear to be a more attractive option because they are clearly capable (in an embryonic environment) of generating all the tissues of the human body. In practice, however, it is extraordinarily difficult to get stem cells of any age "to do what you want them to" in culture.

There are two important counterarguments to the assertion that the therapeutic potential of adult stem cells is less than that of embryonic stem cells because adult cells are "restricted" and therefore unable to generate the full range of mature cell types. First, it is not clear at this point whether adult stem cells are more restricted than their embryonic counterparts. It is important to bear in mind that the field of adult stem cell research is not nearly as advanced as the field of embryonic stem cell research. Scientists have been working on embryonic stem cells for more than a decade, whereas adult stem cells have only been described within the last few years. With few exceptions, adult stem cell research has demonstrated equal or greater promise than embryonic stem cell research at a comparable stage of investigation. Further research may very well prove that it is just as easy to teach an old dog new tricks as it is to train a willful puppy. This would not eliminate the very real problems associated with teaching any dog to do anything useful, but it would remove the justification for "age discrimination" in the realm of stem cells.

The second counterargument is even more fundamental. Even if adult stem cells are unable to generate the full spectrum of cell types found in the body, this very fact may turn out to be a strong scientific and medical advantage. The process of embryonic development is a continuous trade-off between potential and specialization. Embryonic stem cells have the potential to become anything, but are specialized at nothing. For an embryonic cell to specialize, it must make choices that progressively restrict what it can become. The greater the number of steps required to achieve specialization, the greater the scientific challenge it is to reproduce those steps in culture. Our current understanding of embryology is nowhere near advanced enough for scientists to know with confidence that we have gotten all the steps down correctly. If adult stern cells prove to have restricted rather than unlimited potential, this would indicate that adult stem cells have proceeded at least part way toward their final state, thereby reducing the number of steps scientists are required to replicate in culture. The fact that adult stem cell development has been directed by nature rather than by scientists greatly increases our confidence in the normalcy of the cells being generated.

There may well be multiple adult stem cell populations, each capable of forming a different subset of adult tissues, but no one population capable of forming everything. This limitation would make certain scientific enterprises considerably less convenient. However, such a restriction in "developmental potential" would not limit the therapeutic potential of adult stem cells for treatment of disease and injury. Patients rarely go to the doctor needing a full body replacement. If a patient with heart disease can be cured using adult cardiac stem cells, the fact that these "heart restricted" stem cells do not generate kidneys is not a problem for the patient.

The field of stem cell research holds out considerable promise for the treatment of disease and injury, but this promise is not unlimited. There are real, possibly insurmountable, scientific challenges to the use of embryonic stem cells as a medical treatment for disease and injury. In contrast, adult stem cell research holds out nearly equal promise while circumventing the enormous social, ethical, and political issues raised by the use of human embryos for research. There is clearly much work that needs to be done before stem cells of any age can be used as a medical treatment. It seems only practical to put our resources into the approach that is most likely to be successful in the long run. In light of the serious problems associated with embryonic stem cells and the relatively unfettered promise of adult stem cells, there is no compelling scientific argument for the public support of research on human embryos.

Jerusalem Artichoke

It's *not*, though: Anyone can tell you this is
absolutely not an artichoke.
And the *Jerusalem* that it professes?
Some Italian costermonger spoke
about the sunflower—the English heard
not *girasole* but, *Jerusalem*—
and, naming these vile tubers, they conferred
an accidental dignity on them.

People say a flower shows us
certain proof that there's a God;
is *Helianthus tuberosus'*
ugly tuber—sunk in sod—
proof of the devil? If the flower
needs the root, the devil's sent
expressly by a higher power:
Neither thing's an accident.

Deborah Warren

NOTE

1. David Anderson, "The Alchemy of Stem Cell Research," *New York Times*, July 15, 2001, sec. 4, p. 15, col. 1 (Editorial).

26
Locating Convergence: Ethics, Public Policy, and Human Stem Cell Research

A Commissioned Paper for the NBAC

Andrew W. Siegel

INTRODUCTION: BENEFITS AND CONSTRAINTS

The principal moral justification for promoting research with human pluripotent stem cells is that such research has the potential to lead to direct health benefits to individuals suffering from disease. Research that identifies the mechanisms controlling cell differentiation would provide the foundation for directed differentiation of pluripotent stem cells to specific cell types. The ability to direct the differentiation of stem cells would, in turn, advance the development of therapies for repairing injuries and pathological processes. The great promise of human stem cell research inspired thirty-three Nobel laureates to voice their support for the research and to lay down the gauntlet against those who oppose it: "Those who seek to prevent medical advances using stem cells must be held accountable to those who suffer from horrible disease and their families, why such hope should be withheld."[1]

Reprinted from the National Bioethics Advisory Commission (NBAC).

While the invocation of the potential benefits of stem cell research furnishes strong moral grounds for supporting the research, considerations of social utility are not always sufficient to morally justify actions. There are moral constraints on the promotion of the social good. For example, considerations of justice and respect for persons often trump considerations of social utility. Those who oppose research involving the use of stem cells derived from embryos and fetal tissue argue that the research is morally impermissible because it is implicated in the unjust killing of human beings who have the moral status of persons. Opponents of the research maintain that the constraints against killing persons to advance the common good apply to fetuses and embryos. On this view, one cannot approve of a policy supporting such research without holding that elective abortion and the destruction of embryos are morally permissible.

In this paper, I consider whether it is possible to offer a justification for federal support of research with stem cells derived from fetal tissue and embryos that is not premised on a view about the morality of abortion and the destruction of embryos. The aim of seeking such a justification is to provide a foundation for consensus on public policy in the area of stem cell research. I will argue that people with divergent views about the moral status of fetuses and embryos should be able to endorse some research uses of stem cells derived from these sources.

RESEARCH WITH STEM CELLS DERIVED FROM FETAL TISSUE

The ethical acceptability of deriving stem cells from the tissue of aborted fetuses is, for some, closely connected to the morality of elective abortion. For those who believe that abortion is permissible because the fetus has no moral standing, there are no significant moral barriers to research using stem cells derived from fetal tissue. Restrictions that separate decisions to donate fetal tissue from decisions to abort might be thought necessary, but the purpose of the restrictions would be to protect the mother against coercion and exploitation rather than to protect the fetus.

What is less clear is whether one can both morally oppose abortion and support this method of deriving stem cells. A common view in the literature on the ethics of human fetal tissue transplantation research is that we can support the research without assuming that abortion is morally permissible. As long as guidelines are in place that ensure that abortion decisions and procedures are separated from considerations of fetal tissue procurement and use in research, using aborted fetuses for research is no more problematic than using other cadavers donated for scientific and medical purposes.

Opponents of the research use of fetal materials obtained from induced abortions dispute the claim that we can dismiss the relevance of the morality of abortion. They appeal to two grounds in asserting the relevance of the moral question: (1) those who procure and use fetal material from induced abortions are complicit with the abortions which provide the material, and (2) it is impossible to obtain valid informed consent for the use of fetal materials. Each of these claims merits consideration.

Complicity

There are two general ways in which one can maintain that those involved in the research use of stem cells derived from aborted fetuses are complicit with abortion: (1) They bear some causal responsibility for abortions, or (2) they symbolically align themselves with abortion.

Causal Responsibility

A researcher or tissue procurer may bear direct or indirect causal responsibility for abortions. The former kind of responsibility exists where one directly motivates a woman to have an abortion (e.g., by offering financial incentives) or is directly involved in the performance of an abortion from which fetal tissue is procured. There are several measures that can help prevent direct responsibility for abortions:

- A requirement that the consent of women for abortions be obtained prior to requesting or obtaining consent for the donation of fetal tissue
- A requirement that those who seek a woman's consent to donate not discuss fetal tissue donation prior to her decision to donate

- A prohibition against the sale of fetal tissue
- A prohibition against directed donation of fetal tissue
- A separation between tissue procurement personnel and abortion clinic personnel
- A prohibition against any alteration of the timing of or procedures used in an abortion solely for the purpose of obtaining tissue[2]

Those involved in research uses of stem cells derived from fetal tissue would be indirectly responsible for abortions if the perceived benefits (or promise of benefits) of the research contributed to an increase in the number of abortions. Opponents of fetal tissue research argue that it is not realistic to suppose that we can always keep a woman's decision to abort separate from considerations of donating fetal tissue, as many women facing the abortion decision are likely to have already gained knowledge about fetal tissue research through widespread media attention to the issue. The knowledge that having an abortion might promote the common goodwill, opponents argue, tips the balance in favor of going through with an abortion for some of the women who are ambivalent about it. More generally, some also argue that the benefits achieved through the routine use of fetal tissue will further legitimize abortion and result in more socially permissive attitudes and policies concerning abortion.

Although there has been one empirical study examining whether the potential for fetal tissue transplantation is likely to influence abortion decisions, the issue remains largely speculative. The study involved a survey which asked, "If you became pregnant and knew that tissue from the fetus could be used to help someone suffering from Parkinson's disease, would you be more likely to have an abortion?"[3] Twelve percent of the women responded in the affirmative. The authors conclude that the option to donate tissue may influence some women's abortion decisions. The main deficiency of the study, however, is its reliance on a hypothetical that is stripped of the complexities of the actual circumstances a pregnant woman considering abortion might be operating under. As Dorothy Vawter and Karen Gervais have noted: "[G]iven how situation-specific women's abortion decisions are, it is unclear what useful information can be obtained from asking women global hypothetical questions about whether they believe the option to donate would affect their decision to terminate a 'generic' pregnancy sometime in the future."[4]

It is difficult to deny that there is a risk that knowledge of the promise of research on stem cells derived from fetal tissue will play a role in some abortion decisions, even if only very rarely. However, it is not clear J-5 that much moral weight ultimately attaches to this fact. One might be justified in some instances in asserting that "but for" research using fetal tissue a particular woman would not have chosen abortion. But one might assign this kind of causal responsibility to a number of factors which figure into abortion decisions without making ascriptions of complicity. For example, a woman might choose to have an abortion principally because she does not want to slow the advancement of her education and career. She might not have had an abortion in the absence of policies that encourage career development. Yet, we would not think it appropriate to charge those who promote such policies as complicit in her abortion. In both this case and that of research, the risk of abortion is an unintended consequence of a legitimate social policy. The burden on those seeking to end such policies is to show that the risks of harm resulting from the policies outweigh the benefits.[5] This minimally requires evidence of a high probability of a large number of abortions that would not have occurred in the absence of those policies. There is, however, no such evidence at present; nor is there any reason to think that it is forthcoming.

Symbolic Association

Agents can be complicit with wrongful acts for which they are not causally or morally responsible. One such form of complicity arises from an association with wrongdoing that symbolizes an acquiescence in the wrongdoing. As James Burtchaell characterizes it, "It is the sort of association which implies and engenders approbation that creates moral complicity. This situation is detectable when the associate's ability to condemn the activity atrophies."[6] Burtchaell maintains that those involved in research on fetal tissue enter a symbolic alliance with the practice of abortion in benefiting from it.

A common response to this position is that there are numerous circumstances in which persons benefit from immoral acts without tacitly approving of those acts. For example, transplant surgeons and patients may benefit from deaths resulting from murder and drunken driving but nevertheless condemn the wrongful acts.[7] A researcher who benefits from an aborted fetus need not sanction the act of abortion any more than the transplant surgeon who uses the organs of a murder victim sanctions the homicidal act.

This response has not, however, been satisfactory to opponents of fetal tissue research. They maintain that fetal tissue research implicates those involved in a different and far greater evil than is the transplant surgeon in the example above. Unlike drunken driving and murder, abortion is an institutionalized practice in which a certain class of humans (which pro-lifers regard as the moral equivalent of persons) are allowed to be killed. In this respect, some foes of abortion suggest that fetal tissue research is more analogous to research which benefits from victims of the Holocaust.[8]

But whatever one thinks of comparisons between the victims of Nazi crimes and aborted fetuses—and many are understandably outraged by them—one could concede the comparisons without concluding that fetal tissue research is morally problematic. There are, of course, some who believe that those who use data derived from Nazi experiments are morally complicit with those crimes. For example, Seidelman writes: "By giving value to (Nazi) research we are, by implication, supporting Himmler's philosophy that the subjects' lives were 'useless.' This is to argue that, by accepting data derived from their misery we are, *post mortem*, deriving utility from otherwise 'useless' life. Science could thus stand accused of giving greater value to knowledge than to human life itself."[9] But one need not adopt this stance. Instead, one can reasonably hold that the symbolic meaning of scientists' actions must be divined solely from their intentions. As Benjamin Freedman argues:

> A moral universe such as our own must, I think, rely on the authors of their own actions to be primarily responsible for attaching symbolic significance to those actions. . . . [I]n using the Nazi data, physicians and scientists are acting pursuant to their own moral commitment to aid patients and to advance science in the interest of humankind. The use of data is predicated upon that duty, and it is in seeking to fulfill that duty that the symbolic significance of the action must be found.[10]

One could likewise maintain that the symbolic significance of support for research using stem cells derived from fetal tissue lies in the desire to promote public health and save lives. This research is allied with a noble cause, and any taint that might attach from the source of the stem cells arguably diminishes in proportion to the potential good that the research may yield.

One who opposes the use of fetal tissue might reply, however, that those who use Nazi data can symbolically disassociate themselves from the immoral acts that produced the data only because the Nazi regime no longer exists. Those who use fetal tissue cannot —so the argument runs—divorce themselves from the immoral act that produces it because elective abortion is legally protected and largely socially accepted in our society. Opponents of fetal tissue research might seek to generate different intuitions about the moral permissibility of research using fetal tissue by offering a different analogy. Imagine the following: You live in a society where members of a particular racial or ethnic group are regularly killed when they are an unwanted burden. The practice is legally protected and generally accepted. Suppose that biological materials obtained from these individuals subsequent to their deaths are made available for research uses. Would it be morally problematic for researchers to use these materials?

It is likely that many who consider it permissible to use Nazi data would think it problematic to use materials derived from the deceased individuals in the hypothetical scenario. Arguably, what underlies the judgment that the research in the hypothetical is morally problematic is the belief that there is a heightened need to act in ways that protest moral wrongs where those wrongs are currently socially accepted and institutionalized. Attempts to benefit from the moral wrong in these circumstances may be viewed as incompatible with mounting a proper protest.

On the pro-life view, the hypothetical case is analogous to the case of research uses of fetal material. Hence, if we concede (at least for the sake of argument) that the fetus has the moral status of a person, it is not clear that an investigator can engage in fetal tissue research without implicitly sanctioning abortion. However, it is not obvious that this conclusion cuts against a public policy which supports fetal tissue research. Federal funding of research may be problematic where taxpayers who oppose the research would regard themselves as complicit with abortion as a result of the funding. There would, for example, be a basis for concern if fetal tissue research played an important causal role in an increase in the number of elective abortions. But funding research in which investigators are complicit with abortion through their symbolic association with it does not render opponents complicit with abortion. That a person's tax dollars are being used to support fetal tissue research does not imply that he or she acquiesces in abortion, nor does it diminish his or her ability to condemn abortion.

Consent

Many people hold that women should not be allowed to terminate a pregnancy solely for the purpose of donating fetal material. There are two consent requirements that can help insulate the decision to donate from the decision to abort: (1) informed consent for an abortion must be provided prior to an informed consent to donate the fetal tissue, and (2) in the consent process for abortion, there must be no (unsolicited) mention of the possibility of using fetal materials in research and transplantation. The most serious charge against these restrictions on the consent process is that it is disrespectful of the autonomy of women considering abortion to withhold information from them regarding the donation of fetal tissue. Because this information might be important to a woman's abortion decision, the failure to disclose the information would render the consent for the abortion ethically invalid.[11]

There are, however, a number of difficulties with this argument. First, it is not clear that information about the possibility of donation is materially relevant to the abortion decision, since, as discussed above, there is not adequate evidence that the option of donation (where financial incentives and directed donations are prohibited) would ever function as a reason for a woman to abort a fetus. Second, assuming the possibility of donation is materially relevant to some women's abortion decisions, there is an obligation not to disclose the option if it is unethical for women to abort for this purpose. Finally, if clinic personnel are permitted to discuss donation prior to obtaining a woman's consent for abortion, women may be (or feel) pressured to have an abortion, in which case the voluntariness of the consent will be in doubt.[12]

Another problem about consent concerns the matter of who has the moral authority to consent to donate fetal tissue. Some object that, from an ethical standpoint, a woman who chooses abortion forfeits her right to determine the disposition of the dead fetus. Burtchaell, for instance, argues that "the decision to abort, made by the mother, is an act of such violent abandonment of the maternal trusteeship that no further exercise of such responsibility is admissible."[13]

John Robertson argues that this position mistakenly assumes that the disposer of cadaveric remains acts as the guardian or proxy of the deceased. Instead, "a more accurate account of their role is to guard their own feelings and interests in assuring that the remains of kin are treated respectfully."[14] But even if we suppose that a woman does forfeit her moral authority to determine the disposition of her aborted fetus, it is not clear that informed consent is always ethically required for the use of cadaveric remains. The requirement of informed consent in medical practice is largely meant to protect the autonomy of persons. In the absence of any person whose autonomy must be respected, it does not seem that the failure to obtain consent violates anyone's rights.[15]

RESEARCH WITH STEM CELLS DERIVED FROM EMBRYOS

Research with stem cells obtained from human embryos poses moral difficulties that do not arise in the case of fetal tissue. Whereas researchers using fetal tissue are not responsible for the death of the fetus, researchers using stem cells derived from embryos will often be implicated in the destruction of the embryo. Researchers using stem cells derived from embryos are clearly implicated in the destruction of the embryo where they (1) derive the cells themselves, or (2) enlist others to derive the cells. However, there may be circumstances in which opponents of embryo research could not properly deem researchers who use embryonic stem cells complicit with the destruction of embryos. Suppose, for example, that X creates a cell line for his own study and later makes an unsolicited offer to share the cell line with Y so that Y may pursue her own research. Is Y implicated in X's act of destroying the embryo from which the cell line was derived? It does not seem that Y is implicated in X's act. As Robertson argues, it does not appear one can assign causal or moral responsibility for the destruction of an embryo to an investigator where his or her "research plans or actions had no effect on whether the original immoral derivation occurred."[16] Nonetheless, it does seem evident that much research with embryonic stem cells will be morally linked to the derivation of the cells (and the resulting destruction of the embryo), especially in the early stages of the research. Thus, an analysis of the ethics of research with embryonic stem cells, as well as the ethics of the funding of this research, must address the issue of the moral permissibility of destroying embryos.

The Moral Status of Embryos

The moral permissibility of destroying embryos turns principally on the moral status of the embryo. The debate about the moral status of embryos has traditionally revolved around the question of whether the embryo has the same moral status as children and adult humans, with a right to life that cannot be sacrificed for the benefit of society. At one end of the spectrum of positions is the view that the embryo is a mere cluster of cells which has no more moral standing than any other human cells. From this perspective, there are few, if any, limitations on research uses of embryos. At the other end of the spectrum is the view that embryos have the moral status of persons. On this view, research involving the destruction of embryos is absolutely prohibited. An intermediate position is that the embryo merits respect as human life, but not the level of respect accorded persons. Whether research using embryos is acceptable on this account depends upon just how much respect the embryo is thought to deserve.

While the moral permissibility of research using embryonic stem cells turns upon the status of the embryo, the prospects of mediating the standoff between opposing views on the matter are dim. A brief consideration of the competing positions will reveal some of the difficulties of resolving the issue.

The standard move made by those who deny the personhood of embryos is to identify one or more psychological or cognitive capacities that are thought essential to personhood (and a concomitant right to life) but which embryos lack. The capacities most commonly cited include consciousness, self-consciousness, and reasoning.[17] The problem faced by such accounts is that they seem either under- or overinclusive, depending on which capacities are invoked. If one requires self-consciousness or reasoning, most early infants will not satisfy the conditions for personhood. If sentience is regarded as the touchstone of the right to life, then nonhuman animals will also possess this right. Since most of those who reject the personhood of the embryo believe that newborn infants do possess a right to life and animals do not, these capacities cannot generally be accepted as morally distinguishing embryos from other human beings.

Of course, those who reject that embryos have the standing of persons can maintain that the embryo is simply too nascent a form of human life to merit the kind of respect that we accord more developed humans. However, in the absence of an account which decisively identifies the first stage of human development at which destroying human life is morally wrong, one can hold that it is not permissible to destroy embryos.

The fundamental argument of those who oppose the destruction of human embryos is that these embryos are human beings, and as such, have a right to life. The humanity of the embryo is thus thought to confer the status of a person upon it. The problem is that for some, the premise that all human beings have a right to life (i.e., are persons in the moral sense) is not self-evidently true. Indeed, some believe that the premise conflates two categories of "human beings"—namely, beings which belong to the species *homo sapiens*, and beings which belong to the moral community.[18] According to this view, the fact that a being belongs to the species *homo sapiens* is not sufficient to confer on it membership in the moral community. While it is not clear that those who advance this position can establish the point at which human beings first acquire the moral status of persons, those who oppose the destruction of embryos likewise fail to establish that we should ascribe the status of persons to human embryos.

Those constructing public policy on the use of embryos in research would do well to avoid attempting to settle the debate over the moral status of embryos. Ideally, public policy recommendations should be formulated in terms which individuals with opposing views on the status of the embryo can accept. As Thomas Nagel argues, "In a democracy, the aim of procedures of decision should be to secure results that can be acknowledged as legitimate by as wide a portion of the citizenry as possible."[19] Amy Gutmann and Dennis Thompson similarly argue that the construction of public policy on morally controversial matters should involve a "search for significant points of convergence between one's own understanding and those of citizens whose positions, taken in their more comprehensive forms, one must reject."[20]

Locating Convergence

R. Alta Charo suggests an approach for informing policy in this area that seeks to accommodate the interests of individuals who hold conflicting views on the status of the embryo. Charo argues that the issue of moral status can be avoided altogether by addressing the proper limits of embryo research in terms of political philosophy rather than moral philosophy:

> The political analysis entails a change in focus, away from the embryo and the research and toward an ethical balance between the interests of those who oppose destroying embryos in research and those who stand to benefit from the research findings. Thus, the deeper the degree of offense to opponents and the weaker the opportunity for resorting to the political system to impose their vision, the more compelling the benefits must be to justify the funding.[21]

In Charo's view, once we recognize that the substantive conflict among fundamental values surrounding embryo research cannot be resolved in a manner that is satisfactory to all sides, the most promising move is to perform some type of cost-benefit analysis in considering whether to proceed with the research. Thus, one could acknowledge that embryo research will deeply offend many people, but argue that the potential health benefits for this and future generations outweigh the pain experienced by opponents of the research.

It is, however, questionable whether Charo's analysis successfully brackets the moral status issue. One might object that placing the lives of embryos in this kind of utilitarian calculus will only seem appropriate to those who already presuppose that embryos do not have the status of persons. After all, we would expect most of those who believe—or who genuinely allow for the possibility—that embryos have the status of persons, to regard such consequentialist grounds for sacrificing embryos as problematic.

An acceptable political approach must seek to develop public policy around points of convergence in the moral positions of those who disagree about the status of the embryo. Of course, as long as the disagreement is cast strictly as one between those who think the embryo is a person with a right to life and those who think it has little or no moral standing, the quest for convergence will be an elusive one. But there are grounds for supposing that this is a misleading depiction of the conflict. Once this is recognized, it will become clear that there may be sufficient consensus on the status of embryos to justify some research uses of stem cells derived from them.

In his discussion of the abortion debate, Ronald Dworkin maintains that, despite their rhetoric, a large faction of the opposition to abortion does not actually believe that the fetus is a person with a right to life. This is revealed, he claims, through a consideration of the exceptions that they permit to their proposed prohibitions on abortion:

> It is a very common view, for example, that abortion should be permitted when necessary to save the mother's life. Yet this exception is also inconsistent with any belief that a fetus is a person with a right to live. Some people say that in this case a mother is justified in aborting a fetus as a matter of self-defense; but any safe abortion is carried out by someone else—a doctor—and very few people believe that it is morally justifiable for a third party, even a doctor, to kill one innocent person to save another.[22]

It can further be argued that one who regards the fetus as a person would seem to have no reason to privilege the mother's life over the fetus's life. Indeed, inasmuch as the mother typically bears some responsibility for putting the fetus in the position in which it needs her aid, one who regards the fetus as a person might well hold that the fetus's life should be privileged over the mother's life.

Some abortion conservatives further hold that abortion is morally permissible when a pregnancy is the result of rape or incest. Yet, as Dworkin comments, "it would be contradictory to insist that a fetus has a right to live that is strong enough to justify prohibiting abortion even when childbirth would ruin a mother's or a family's life, but that ceases to exist when the pregnancy is the result of a sexual crime of which the fetus is, of course, wholly innocent."[23]

The importance of these exceptions in the context of research uses of embryos is that they suggest we can identify some common ground between liberal and conservative views on the permissibility of destroying embryos. Conservatives who allow these exceptions implicitly hold with liberals that very early forms of human life can sometimes be sacrificed to promote the interests of other humans. While liberals and conservatives disagree about the range of ends for which embryonic or fetal life can ethically be sacrificed, they should be able to reach some consensus. Conservatives who accept that killing a fetus is permissible where it is necessary to save a pregnant woman or spare a rape victim additional trauma could agree with liberals that it is also permissible to destroy embryos where it is necessary to save people or prevent extreme suffering. Of course, the cases are different inasmuch as the existence of the fetus directly conflicts with the pregnant woman's interests, while a particular *ex utero* embryo does not threaten anyone's interests. But this distinction does not bear substantial moral weight. It is the implicit attribution of greater value to the interests of adult humans over any interests the fetus may have that informs the judgment that it is permissible to kill it in the cases at issue. Thus, the following would seem a reasonable formulation of a position on embryo research that liberals and conservatives could agree upon: *Research that involves the destruction of embryos is permissible where there is good reason to believe that it is necessary to cure life-threatening or severely debilitating diseases.*

Given the great promise of stem cell research for saving lives and alleviating suffering, this position would appear to permit research uses of stem cells derived from embryos. Some might object, however, that the benefits of the research are too uncertain to justify a comparison with the abortion exceptions. But the lower probability of benefits from research uses of embryos is balanced by a much higher ratio of potential lives saved for lives lost. Another objection is that it is unnecessary to use embryos for stem cell research because there are alternative means of obtaining stem cells. This reflects an important concern. The derivation of stem cells from embryos is justifiable only if there are no less morally problematic alternatives for advancing the research. At present, there appear to be strong scientific reasons for using embryos. But this is a matter that must continually be revisited as science advances.

It is important to further note that the abortion exceptions that serve as the basis for the type of convergence identified above are exceptions to the law banning federal funding for abortions.[24] Thus, federal funding of research uses of embryos within such limitations as those specified above can reasonably be viewed as consistent with current federal funding practices in the abortion context.

There is, however, one further qualification that must be made here. The justification developed above for federal funding of research that involves the destruction of embryos does not appear to extend to research in which embryos are created expressly for research purposes. While there is wide agreement that it is acceptable to produce embryos for in vitro fertilization (IVF), there is much controversy about the moral permissibility of creating embryos for research. Indeed, many who approve of research using embryos remaining from IVF treatments oppose the creation of embryos solely for research purposes.[25] The main basis for the distinction is that, whereas embryos created for procreative purposes are originally viewed as potential children, embryos created for research are meant to be treated as mere objects of study from the outset. Some dispute that there is a morally relevant distinction between the use of spare IVF embryos and embryos created for research.[26] But, regardless of which view is more sound, the foundation for agreement on embryo research identified above does not appear to support the use of embryos that are created expressly for research purposes. This limitation on federal funding of stem cell research could ultimately prove to be a significant impediment to the advancement of the research. However, it should come as little surprise that parties on both sides of the issue may incur substantial costs when constructing public policy around points of convergence between competing visions of the meaning and value of life.

NOTES

1. Letter to Congress and President Clinton, March 4, 1999.
2. National Institutes of Health (NIH), *Report of the Fetal Tissue Transplantation Research Panel* (Bethesda, Md.: NIH, 1988).
3. D. K. Martin, et al., "Fetal Tissue Transplantation and Abortion Decisions: A Survey of Urban Women," *Canadian Medical Association Journal* 153, no. 5 (1995): 548.
4. D. E. Vawter and K. G. Gervais, "Commentary on Abortion and Fetal Tissue Transplantation," *IRB: A Review of Human Subjects Research* 15, no. 3 (1993): 5.
5. J. F. Childress, "Ethics, Public Policy, and Human Fetal Tissue Transplantation Research," *Kennedy Institute of Ethics Journal* 1, no. 2 (1991): 93–121.

6. J. T. Burtchaell, "The Use of Aborted Fetal Tissue in Research: A Rebuttal," *IRB: A Review of Human Subjects Research* 11, no. 2 (1989): 9–12.

7. J. Robertson, "Fetal Tissue Transplant Research Is Ethical," *IRB: A Review of Human Subjects Research* 6, no. 10 (1988): 5–8; D. Vawter et al., "The Use of Human Fetal Tissue: Scientific, Ethical, and Policy Concerns," *International Journal of Bioethics* 2, no. 3 (1991): 189–96. Dorothy Vawter et al., *The Use of Human Fetal Tissue: Scientific, Ethical, and Policy Concerns* (Minneapolis: Center for Biomedical Ethics, University of Minnesota, 1990).

8. J. Bopp Jr., "Fetal Tissue Transplantation and Moral Complicity with Induced Abortion," in *The Fetal Tissue Issue: Medical and Ethical Aspects*, ed. P. Cataldo and A. Moracszewski (Braintree, Mass.: Pope John Center, 1994), pp. 6–79.

9. W. E. Seidelman, "Mengele Medicus: Medicine's Nazi Heritage," *Milbank Quarterly* 66 (1988): 221–39.

10. B. Freedman, "Moral Analysis and the Use of Nazi Experimental Results," in *When Medicine Went Mad*, ed. A. L. Caplan (Totowa, N.J.: Humana Press, 1992), p. 151.

11. D. K. Martin, "Abortion and Fetal Tissue Transplantation," *IRB: A Review of Human Subjects Research* 15, no. 3 (1993): 1–3.

12. Vawter and Gervais, "Commentary on Abortion and Fetal Tissue Transplantation."

13. Burtchaell, "The Use of Aborted Fetal Tissue in Research: A Rebuttal."

14. Robertson, "Fetal Tissue Transplant Research Is Ethical," p. 6.

15. D. G. Jones, "Fetal Neural Transplantation: Placing the Ethical Debate within the Context of Society's Use of Human Material," *Bioethics* 3, no. 1 (1991): 22–43.

16. J. Robertson, "Ethics and Policy in Embryonic Stem Cell Research," *Kennedy Institute of Ethics Journal* 2, no. 9 (1999): 109–36.

17. M. A. Warren, "On the Moral and Legal Status of Abortion," *Monist* 57 (1973): 43–61; M. Tooley, *Abortion and Infanticide* (New York: Oxford University Press, 1983); J. Feinberg, "Abortion," in *Matters of Life and Death*, ed. T. Regan (New York: Random House, 1986), pp. 256–93.

18. Warren, "On the Moral and Legal Status of Abortion."

19. T. Nagel, "Moral Epistemology," in *Society's Choices*, ed. R. E. Bulger, E. M. Bobby, and H. V. Fineberg (Washington, D.C.: National Academy Press, 1995), pp. 201–14.

20. A. Gutmann and D. Thompson, *Democracy and Disagreement* (Cambridge, Mass.: Belknap Press, 1996).

21. R. A. Charo, "The Hunting of the Snark: The Moral Status of Embryos, Right-to-Lifers, and Third World Women," *Stanford Law and Policy Review* 6 (1995): 11–27.

22. R. Dworkin, *Life's Dominion: An Argument about Abortion, Euthanasia, and Individual Freedom* (New York: Vintage, 1994), p. 32.

23. Ibid.

24. Senate Appropriations Subcommittee on Labor, Health and Human Services, and Education, *Title V*, 112 Stat. 3681-385, Sec. 509(a) (1) and (2).

25. G. Annas, A. L. Caplan, and S. Elias, "The Politics of Human-Embryo Research: Avoiding Ethical Gridlock," *New England Journal of Medicine* 334, no. 20 (1996): 1329–32.

26. J. Harris, *Clones, Gene, and Immortality: Ethics and the Genetic Revolution* (New York: Oxford University Press, 1998).

27
Stem Cells

Public Policy and Ethics

Cindy R. Towns and D. Gareth Jones

INTRODUCTION

Human embryonic stem (ES) cells burst into the limelight in 1998, with their first successful derivation. The attention they have subsequently received, on account of their potential to alleviate a range of debilitating illnesses, and give rise to a new genre of medical therapies, has been bewildering.[1] These positive vistas have been counterbalanced by a welter of concerns, ranging from the ever-present ethical dilemmas precipitated by the moral status of the human embryo, to a confusing array of conflicting claims regarding the scientific superiority of adult stem cell sources. Unfortunately, the calibre of ethical debate has been detrimentally affected by failure to appreciate the subtleties of the scientific background.

For many the least controversial course of action would be to use adult stem cells, and so the pressures on scientists to emerge with evidence demonstrating that their potential is at least as great as that

---

Reprinted from *New Zealand Bioethics Journal* 5, no. 1 (2004): 22–28.

of ES cells are formidable. Such a course of action is appealing to some since it appears to circumvent the ethical problems raised by using human embryos as the source of the stem cells. Unfortunately, scientific uncertainty abounds in this very young field, where core concepts are being refined almost daily. Consequently, it is of concern when policymakers and governments, in addition to ethicists, philosophers, and theologians, demand definitive scientific answers in this rapidly changing terrain. These demands demonstrate a profound misunderstanding of the intimate interrelationship between science, ethics, and public policy. The results of this misunderstanding are clearly seen when the manner in which societies regulate stem cell technology is examined.

THE SCIENCE OF STEM CELLS

It is now well recognized that stem cells are unspecialized cells, which have the ability to renew themselves indefinitely, and under appropriate conditions can give rise to a variety of mature cell types in the human body. They have multiple sources, ranging from embryos (the inner cell mass [ICM] of blastocysts), to umbilical cord blood, fetal tissues, and a variety of adult tissues. For the sake of simplicity stem cells from all sources other than embryos are termed adult stem cells (as opposed to ES cells). While this distinction is a fundamental one, especially in ethical debate, it is less clear than frequently thought, since the identification of stem cells is far from cut-and-dried, and depends to some extent upon the environment. Indeed, there appears to be a dynamic relationship between stem cells and their immediate microenvironment—the stem cell niche.[2]

This niche is characterized by numerous external signals, including those derived from chemical factors, cell-cell interactions, and relationships between cells and the surrounding tissue.[3] These, in their various ways, all have an impact on stem cells, because they affect the precise directions in which they subsequently develop. It is also interesting to note that stem cells taken out of their original niche and exposed to an entirely new environment can potentially differentiate into the cell types typical of that new environment.[4] In other words, stem cells demonstrate considerable plasticity.

While this is a fascinating finding, it would be foolhardy to jump to the conclusion that it renders the use of ES cells unnecessary. There are a number of scientific reasons for this.[5] Even though there are a few confirmed reports of truly pluripotential adult human stem cells,[6] what is required is far more understanding of the fundamental biological issues raised by this research. Scientifically, therefore, research with both adult and embryonic sources should continue, bearing in mind that adult stem cells are more problematic than their embryonic counterparts.

In light of this evaluation, considerable care should be employed in advocating, on allegedly scientific grounds, the advantages of adult over embryonic cells as the source of replacement tissues. In other words, it is shortsighted to attempt to circumvent discussion of the moral status of the blastocyst by concentrating on the scientific potential of adult stem cells alone. What is even more important is that ethical debate and policy regulations need to take into account whether the blastocysts in question are located within an environment appropriate to their further development. If the environment is not conducive to such development, this may have considerable ethical implications for research aimed at isolating ES cells from such blastocysts. Unfortunately, up to the present, the role of the environment has failed to feature in debate on policy development.

PROVIDING A CONTEXT FOR BLASTOCYSTS AND STEM CELLS

This confronts us with the notion of the importance of both the micro- and macro-environments on the identification and plasticity of stem cells. These have implications for notions of totipotency (capability of forming a new individual) and pluripotency (capability of creating all an individual's cell lines but not the individual itself). This distinction is more limited than frequently thought, since it fails to take account of the environment, a factor relevant to our assessment of both the blastocyst (the early embryo at 5–7 days gestation) and ES cells.

Before any cells can be considered to be totipotent, it has to be established that they are capable of giving rise to all three germs layers and, therefore, to all the major organ and tissue types of the individual. This requires the presence, not only of the ICM from which the germ layers and hence future individual are formed, but also of the trophectoderm cells which give rise to the layers of the placenta. These extraembryonic tissues are a crucial source of signaling molecules and must function optimally for the differentiation of both embryonic somatic cells and for the establishment of germ lines.[7] In other words, the trophectoderm as well as the ICM cells are required to establish totipotency.[8] However, if a viable fetus is to result, totipotency also requires successful implantation and development within a woman's uterus.

In the absence of all these conditions ES cells are merely pluripotent, possessing the capacity to create all the cell lines of the fetus, but not the fetus itself. In the laboratory environment they are incapable of totipotency, since they have been removed from the context of the trophectoderm, let alone that of the uterus.[9] It is inaccurate, therefore, to view ES cells as totipotent.[10]

What emerges from these considerations is that, within the laboratory environment, blastocysts are "potentially totipotent" rather than "actually totipotent."[11] In this, they stand in stark contrast to their counterparts within a woman's body. It is unfortunate that ethical debate has concentrated almost exclusively upon blastocysts (embryos) as discrete autonomous entities, as though their potential to become future individuals can be realized regardless of environmental considerations. Since a notion such as totipotency is a function of the environment both at the microscopic and macroscopic levels, the consequences of context for moral debate are considerable.

We are now in a position to assess some of the policy issues encountered in regulating stem cells. It will emerge that some of the above scientific points play little if any part in the formulation of policies. We shall return to this in the final section.

OVERVIEW OF PUBLIC POLICY

In an extremely helpful review of regulatory practices, Knowles[12] divided countries into three groups: embryo research is prohibited; in vitro fertilization (IVF) embryos can be used for human ES cell research; embryos can be explicitly created for human ES cell research by fertilization and/or somatic cell nuclear transfer (SCNT or "research cloning"). In discussing the variation in regulations, Knowles[13] considers that in Europe the more liberal regulations are found in countries with Protestant religious traditions, and the more conservative or restrictive ones in Catholic countries. However, this simple distinction is beginning to break down with some of the traditional Catholic countries, like Spain and France, beginning to allow research on human embryos. It is also of note that the Protestant-Catholic distinction does not apply in the United States.

Knowles[14] regards a number of Asian countries as having lenient regulatory provisions and suggests, along with Gershon,[15] that in Singapore at least this may be due to the significant commercial potential of stem cell technologies. No hint is given as to any cultural or religious reasons underlying these liberal regulations. By contrast, in Iran and other Islamic countries, Knowles[16] recognizes that, by placing ensoulment of the fetus at around 120 days, the use of surplus IVF embryos in research is permissible within a Muslim context. Israel's liberal regulations are put down to the government's enthusiasm for the commercial potential of stem cell research, although Jewish theological stances on early development may prove equally important.

Useful as this summary of current (and rapidly changing) regulations is, it leaves unanswered pertinent questions regarding the nature of the ethical, theological, and cultural debate behind the regulations themselves. Even this brief summary highlights the inadequacy of generalizations, and provides no hint on the consistency or otherwise of the ethical arguments employed to back up the various positions. Use of the designations conservative and liberal are also unhelpful, since they are far too crude to provide substantive insight into individual positions. Further, the use of just three categories ignores a major theme in much current debate—the presence of a time element in some of the regulations.

FOUR UNDERLYING POSITIONS

We argue that four regulatory positions can be identified. These vary from position A, the prohibition of all embryo research, to position D, the creation of human embryos specifically for research—encompassing both fertilization and SCNT. In addition, there are two intermediate positions. Of these, position B confines the use of ES cells to those currently in existence, in that they were extracted prior to some specified date, thereby prohibiting the extraction of ES cells and the utilization of ES cells derived in the future. Position C allows for the use and ongoing isolation of ES cells from surplus IVF embryos.

Of these four positions, we shall concentrate on position B, which has been adopted by the United States, Germany, and Australia (with subtle differences between them). This is a particularly interesting position since it can be regarded as a classic compromise stance, with its dual aims of purportedly protecting the human embryo, while also encouraging some scientific research on human ES cells.

Current legislation in Germany prohibits any research on embryos that leads to their destruction. However, the import of existing cell lines (those extracted prior to January 1, 2002) derived from surplus IVF embryos is allowed, provided that the intended research has clinical goals unachievable by other means.[17] Australia allows embryonic stem cell research to be conducted as long as it is on existing stem cell lines derived from embryos surplus to IVF requirements, and created before April 5, 2002.[18]

While the United States government has no legislation as such to regulate embryonic stem cell research, federal funding is restricted to research that uses ES cells derived from surplus IVF embryos prior to August 9, 2001. These guidelines prohibit the ongoing extraction of stem cells, and the creation of embryos for research.[19] In contrast, there are no federal restrictions on privately funded research, which is subject to state laws alone. States vary in their guidelines, with some, such as California and New Jersey, encouraging both embryonic and adult stem cell research.

POSITION B—AN ASSESSMENT

Position B is an alluring one, since in one stroke it gives the impression of placating both sides of an exceedingly contentious argument. Research can continue, with at least a modicum of governmental support, although the severity of the scientific limitations under which it is being carried out will not be evident to most people. What is more, those advocating protection of human embryos can feel that their case has been supported, by the prevention of the destruction of any more embryos for research (and possibly therapeutic) purposes. However, we view this way forward as little more than a political construct, with a very unconvincing ethical basis.

The striking feature of this position is that, while it is based on the moral unacceptability of embryo destruction, it allows the use of existing cell lines. Since these have been obtained through the destruction of embryos, the policy implicitly accepts the legitimacy of embryo destruction, albeit in the past. If this were not the case, no research of any description utilizing human embryos or ES cell lines would be tolerated. Position A, with its prohibition of any such research, would be the stance of choice. On the other hand, an unwillingness to move to position C, permitting the extraction and utilization of ES cells, demonstrates that the destruction of human embryos is deplored. Position B represents an uneasy compromise, made possible only by accepting the use of "ethically" tainted/ unethically "derived" material.

This is not a new phenomenon, but is one that has been debated in various forums since the events of the 1930s and 1940s in Germany. Examples include the use of anatomical specimens or data derived from unethical experiments on human beings during the Nazi era, or more recently, the use of neural tissue obtained from aborted fetuses for the treatment of patients with Parkinson's disease. These are examples of instances that raise the concept of moral complicity.

These are relevant for the ES cell debate since, if it is contended that the extraction of stem cells from embryos is unethical (necessitating, as it does, the destruction of embryos), moral complicity demands that any subsequent use of the extracted tissue will also be unethical. There is an indissoluble ethical link between the two, with the origin of the material affecting whatever may later be done with it. Its unethical origins in embryo destruction ensure that any resulting ES cell research is mired in maleficence. Is there any way around this?

There may be. A long-established position is to argue that the unethically derived material from the Nazi era can be used ethically on condition that the wrongs are acknowledged and recurrence is prevented.[20] A similar argument, but omitting these provisos, has been employed to justify the use of aborted fetal tissue in fetal neural grafting.[21] The German, Australian, and United States regulations provide parallels with this approach, since although they imply moral wrongdoing, they also aim to legitimize restricted use of ES cells.

However, if moral complicity is accepted in its entirety, no separation exists between the embryo destruction and the ES cell research; the latter is as unethical as the former. Prohibition, as in position A, is the only acceptable path. On the other hand, if a separation can be forged between the two, and if the unethical nature of the embryo destruction is acknowledged with efforts being made to prevent further embryo destruction, it may prove possible to allow limited research on embryos. This is what the German, Australian, and United States guidelines set out to accomplish, by confining research to embryos or stem cell lines already in existence (position B). The prohibition of any future ES cell extraction means that any further destruction of embryos is outlawed. In a very neat way, past wrongdoing is acknowledged and future wrongdoing is prevented. But, in our view, there is a problem.

The prevention of future embryo destruction, which lies at the heart of this response to moral complicity, fails miserably. And this is for one major reason. Embryo destruction is accepted in all these countries as part of their extensive IVF programs. All such programs are responsible for the production of surplus embryos, most of which will be discarded and hence destroyed. Given the frequency of this practice, what is the rationale for basing stem cell regulations on the premise that embryo destruction is unacceptable? The linkage between the two situations is inescapable, leading to the ironic conclusion that restrictive ES cell guidelines do nothing other than prevent research on embryos that are slated for destruction.

In the case of the United States guidelines an additional consideration applies. The intense debates over these guidelines, and the Herculean efforts to protect the human embryo have limited applicability, that is, to the public sector. This is the domain of federally funded research. The private sector, where so much of the research is being undertaken, is exempt, and researchers here are free to use human embryos for research purposes, and therefore extract ES cells. Consequently, in the United States, despite the vociferous high-level political machinations that are undertaken, human embryos will continue to be destroyed in both IVF programs and in privately funded ES cell research. The ethical dissonance is notable.

SURPLUS EMBRYOS AS A SOURCE OF EMBRYONIC STEM CELLS

There are additional issues to be considered. The model provided by neural grafting in Parkinson's disease has already been alluded to, because similar arguments occur here with respect to moral complicity. What is significant about the neural grafting model is that complicity in the original abortion is avoided because consent to abort the fetus is given on the grounds of maternal or fetal welfare. Quite clearly, the reasons for the abortion and the manner in which it is carried out have nothing to do with the research or therapy on the patients with Parkinson's disease. Indeed, the ethical parameters ensure that there is complete separation in practice between the two.[22] What lessons can be learned from this for ES cell research?

At first glance there is a problem, since the separation cannot be nearly as distinct in ES cell research. This is because the process of acquiring ES cells actually destroys the embryo.[23] The parallel in neural grafting would be if the process of acquiring the neural tissue from the fetus killed the fetus, whereas in practice it is the abortion that does this.

However, this ignores the source of the embryos. ES cells can either be obtained from embryos created specifically for research purposes, or from embryos surplus to IVF programs.[24] The objection raised above only applies to embryos deliberately created for research; the research kills embryos that would not have existed if it were not for this particular research program. By contrast, quite a different situation holds when the embryos in question are surplus to the requirements of IVF treatment. Their existence is independent of this or any other research program, and they will eventually be destroyed regardless of any research interests. A decision to use such embryos for research purposes is completely separate from a decision to discard them, because they are surplus to the reproductive needs of the couple from whom they were obtained.

What are the reasons against using embryos from IVF sources for research purposes, and in particular for extracting ES cells? There appear to be no convincing ethical reasons against this. Nevertheless, what we have seen is that some countries have adopted an intermediate position with built-in time lines (position B). Surprisingly, this position does not seem to stem from a clearly formulated ethical base, but amounts to a desire to protect the human embryo while also allowing the continuation of potentially important research. Unfortunately, it fails on both counts. Its lack of a convincing ethical base opens it to criticism from the conservative side on the grounds that it fails adequately to protect human embryos. Alongside this, the restriction of scientific research to a set number and type of embryos or stem cell lines hampers the dimensions of this research.[25] Such restrictions may have profound implications for scientific understanding and the development of scientific concepts, and even for therapeutic applications. In other words, it is a compromise that is fundamentally flawed on both ethical and scientific counts.

In terms of the principles outlined, a more consistent approach for position B countries would be to adopt position C regulations. This would allow them to hold a protective view of the human embryo, within the framework of a more consistent ethical stance. This is because ES cell research is limited to surplus embryos from IVF programs, with a procedural separation between the initial decision to discard embryos and the subsequent decision to donate them for research. This allows both the utilization and extraction of new ES cells, and eliminates arbitrary time limits on extraction. This is not an ideal solution, but in terms of the principles so far outlined, it is a far more satisfactory one than that enshrined in position B.

WAYS AHEAD

From this account it emerges that, in scientific terms, there is a need to experiment upon ES cells as well as adult stem cells. While any country is at liberty to prohibit such research, the state of the science points to its necessity if optimum progress is to be made in understanding the scientific and clinical potential of stem cells. For all other countries regulations are needed to cover their use. It has also emerged in the preceding sections that environmental considerations are integral to ethical debate. How are these considerations to be taken note of in policy regulations?

We have already argued that position B cannot be supported on ethical grounds, even in countries that do not wish to destroy further embryos. Hence, even for them, position C appears to be an interim position of choice. But what about the creation of embryos for research purposes, either by fertilization or SCNT, and the move to position D?

The environmental argument demonstrates that blastocysts in the laboratory are not totipotent, regardless of their source. They have no chance of becoming individuals unless transferred to a woman's uterus. Since this is not the intention when dealing with surplus IVF embryos or those created for research purposes, research on these embryos will not destroy embryos that would have become fully fledged humans. Even their pluripotency will be lost if they are not used for research or therapy.

Even if embryos (blastocysts) that could never realize their totipotency are considered to be persons, they will never realize that personhood if they exist as surplus IVF embryos or embryos created specifically for research purposes. The differences between these two groups appear minimal, in that they reside in an environment that precludes the realization of human development. Surplus IVF embryos will be destroyed once the need to produce children has passed, regardless of the possibility of being utilized in ES cell research. If used in the latter manner they may contribute to improvements in therapy. Embryos created for specifically research purposes will be destroyed in an attempt to benefit others through improvements in therapy. The move to position D (as found in the United Kingdom) may be the most consistent position ethically and the most advantageous scientifically.

NOTES

1. C. Towns and D. G. Jones, "Stem Cells, Embryos and the Environment: A Context for Both Science and Ethics," *Journal of Medical Ethics* 30 (August 2004): 410–13.

2. F. M. Watt and B. L. Hogan, "Out of Eden: Stem Cells and Their Niches," *Science* 287 (2000): 1427–30.

3. Ibid.

4. P. Wu et al., "Region-Specific Generation of Cholinergic Neurons from Fetal Human Neural Stem Cells Grafted in Adult Rat," *Nature Neuroscience* 5 (2002): 1271–78.

5. Towns and Jones "Stem Cells, Embryos and the Environment: A Context for Both Science and Ethics."

6. National Institutes of Health, *Stem Cells: Scientific Progress and Future Research Directions* (Bethesda, Md.: Department of Health and Human Services, 2001), http://www.nih.gov/news/stemcell/scireport .htm; Committee on the Biological and Biomedical Applications of Stem Cell Research, *Stem Cells and the Future of Regenerative Medicine* (Washington D.C.: National Academy Press, 2002).

7. M. A. Surani, "Reprogramming of Genome Function through Epigenetic Inheritance," *Nature* 414 (2001): 122–28.

8. D. G. Jones and B. Telfer, "Before I Was an Embryo, I Was a Preembryo: Or Was I?" *Bioethics* 9 (1995): 32–49.

9. Towns and Jones "Stem Cells, Embryos and the Environment: A Context for Both Science and Ethics."

10. J. L. Abkowitz, "Can Human Hematopoietic Stem Cells Become Skin, Gut, or Liver Cells?" *Nature Medicine* 8 (2002): 213–14.

11. Towns and Jones "Stem Cells, Embryos and the Environment: A Context for Both Science and Ethics."

12. L. R. Knowles, "A Regulatory Patchwork—Human ES Cell Research Oversight," *Nature Biotechnology* 22 (2004): 157–63.

13. Ibid.

14. Ibid.

15. D. Gershon, "Complex Political, Ethical and Legal Issues Surround Research on Human Embryonic Stem Cells," *Nature* 422 (2003): 928–29.

16. Knowles, "A Regulatory Patchwork—Human ES Cell Research Oversight."

17. Deutscher Bundestag, *Entwurf eines Gesetzes zur Sicherstellung des Embryonenschutzes im Zusammenhang mit Einfuhr und Verwendung menschlicher embryonaler Stammzellen (Stammzellgesetz—StZG)* (Berlin: Bundestagsdrucksache, 2002).

18. *Research Involving Human Embryos Act 2002* (Canberra: Parliament of Australia, 2002).

19. National Institutes of Health, *Stem Cell Information: Pending Stem Cell Research and Cloning Legislation* (Bethesda Md.: Department of Health and Human Services, 2003), http://stemcells.nih.gov/fedPolicy/NIHFed Policy.asp.

20. M. Weitzman, "The Ethics of Using Nazi Medical Data: A Jewish Perspective," *Second Opinion* 14 (1990): 27–38.

21. D. G. Jones, "Fetal Neural Transplantation: Placing the Ethical Debate within the Context of Society's Use of Human Material," *Bioethics* 5 (1991): 23–43.

22. Ibid.

23. R. M. Doerflinger, "The Ethics of Funding Embryonic Stem Cell Research: A Catholic Viewpoint," *Kennedy Institute of Ethics* 9 (1999): 137–50.

24. J. A. Thomson et al., "Embryonic Stem Cell Lines Derived from Human Blastocysts," *Science* 282 (1998): 1145–47; B. E. Reubinoff et al., "Embryonic Stem Cell Lines from Human Blastocysts: Somatic Differentiation In Vitro," *Nature Biotechnology* 18 (2000): 399–404.

25. L. Roccanova, P. Ramphal, and P. Rappa, "Mutation in Embryonic Stem Cells," *Science* 292 (2001): 438–40; D. Kennedy, "Stem Cells: Still Here, Still Waiting," *Science* 300 (2003): 865.

28
Ethical Consistency in Embryonic Stem Cell Research

Simon Clarke

Embryonic stem cell research is controversial because it promises potentially large therapeutic benefits, but extracting stem cells causes the destruction of the embryo, and the moral status of embryos is controversial. Some people think they are persons or potential persons, or that destruction of embryos is for some other reason wrong.

There are four positions government could take on the issue, which Jones and Towns helpfully distinguish:

A. prohibition of all embryo research.
B. confine the use of embryonic stem cells to those currently in existence—extracted prior to some specified date. This approach prohibits the extraction of ES cells, and the utilisation of ES cells derived in the future.
C. allow for the use and ongoing isolation of embryonic stem cells from surplus IVF embryos.

Comments on Gareth Jones and Cindy Towns, "Stem Cells: Public Policy and Ethics." Reprinted from *New Zealand Bioethics Journal* 5, no. 1 (February 2004).

D. *laissez-faire* position—the creation of human embryos specifically for research.

The authors argue against position B. If the countries which take this position—United States, Australia, and Germany—do so on the grounds that embryo destruction is wrong, there is an inconsistency in that these countries also permit IVF programs to help infertile couples have children, programs which also involve the destruction of embryos—those "spare" embryos left over from the IVF procedures. Ethical consistency therefore requires moving from B to C.

I agree with this argument but wish to suggest that the argument may be pushed even further. Ethical consistency requires not only a move from B to C, but also from C to D. Note that accepting B and C implies that it is permissible to sometimes destroy embryos, *even if embryos have some moral status that makes it usually wrong to kill them.* But if so, then why is it not also permissible to create and destroy them for research, as D would let us do? Is there a rationale for C that can stop the move to D?

Consider three possible rationales:

1. Some might think the following moral principle holds: it is wrong to (create and) kill one person to save another. We do not usually believe that it is okay to kill some to save another. D clearly violates this principle, but it might be thought that C does not. In C, the embryos are not destroyed for stem cell research, they are being destroyed anyway so we may as well use them for stem cell research.

But while C does not kill to save another, it does something that must also be considered wrong by those who hold this moral principle: since spare embryos from IVF are destroyed, even putting aside the benefits that might be got from using them for stem cell research, the procedure kills some for the benefit of others—the benefit being helping a childless couple have a child. In fact, important as relieving infertility is, surely this benefit is less important than the benefits stem cell research may lead to: saving people's lives and drastically increasing the quality of some people's lives. So, if anything, D is more justified than C: both involve the destruction of embryos, but the reason in D for doing so is stronger than that in C.

2. Another possible way of distinguishing D from C is to pay attention to the fact that D requires intentionally killing a person,

which many people think is always wrong. With position C, on the other hand, the intention is not to kill but to help a couple have a child. Death may be foreseeable since it is known that spare embryos will be destroyed, but it is unintended. (This argument appeals to the "doctrine of double effect" that is much discussed in bioethics.)

But this is unconvincing. Why does D necessarily involve the intention to kill? The intention could be to develop therapies to help others. Killing is foreseen, perhaps even foreseen as inevitable, but an unintended consequence. This may strike many as ridiculous: if it is known all along that embryos will be destroyed (because they are created for stem cell research), maintaining that there is no intention to kill is just a "head in the sand" position. But surely the same can be said for C; it is known all along that spare embryos from IVF will be destroyed (whether or not they are used for stem cell research), so there is just as much intention to kill here as in D.

3. The third possible rationale for C over D appeals to political reality rather than ethical principle. Position C is a reasonable compromise between the two values of protecting embryos on the one hand and advancing research on the other. Position D on the other hand, goes too much in favor of the first value. While ethical principle cannot justify C, it is good public policy.

But C is not the only possible compromise, nor is it clear that it would be the most acceptable to each side. It fails to give either side what they want—spare embryos from IVF will be destroyed which conservatives won't like; and law will prohibit creating embryos for stem cell research, which liberal researchers won't like. And both are forced to accept a policy that is ethically inconsistent (from either point of view)—we should not underestimate the discomfort people feel in the knowledge that policy is inconsistent.

Here is a different compromise: position D, but with strong encouragements to look for alternative ways of achieving the same goals without destroying embryos. These could take the form of financial incentives, tax breaks, and so on, but no legal prohibitions on creating embryos for sole purpose of stem cell research. This I think is a better compromise: it allows for the advancement of research, but also gives strong symbolic support to protecting embryos. And it provides for the goal of protecting embryos and is ethically consistent.

None of these three rationales succeed. Hence, ethical consistency pushes us all the way from B to D. We must therefore choose between D or A. The choices are stark: either we accept the laissez-faire position of permitting creation of embryos for stem cell research, or we must reject all embryonic stem cell research (and prohibit the destruction of "spare" embryos from IVF programs).

Further Reading

As you will have realized, stem cell research is a rapidly forward moving field, scientifically, medically, and politically. To keep yourself abreast, you should be reading at least one of the major newspapers like the *New York Times* or the *Washington Post* (whose treatments of the topic tend to be truly excellent), as well as looking at some of the more popular science magazines like the *Scientific American* and (in Britain) the *New Scientist*. There are some excellent articles on the issue in the *Hastings Center Report* as well as the *Cambridge Quarterly of Healthcare Ethics*. The two weekly more professional science magazines, *Science* in America and *Nature* in Britain, also carry much good and pertinent material. A good collection of rather technical papers dealing with the science is *Stem Cell Biology*, edited by Daniel R. Marshak, Richard L. Gardner, and David Gottlieb (Cold Spring Harbor Laboratory, 2001), and an excellent collection of papers dealing with more general issues is *The Human Embryonic Stem Cell Debate (Basic Bioethics)*, edited by Suzanne Holland, Karen Lebacqz, and Laurie Zoloth (MIT Press, 2001). Additionally there is the more recent *Stem Cell Now*, by Christopher Thomas Scott (Pi Press, 2005) and the *Stem Cell Handbook*, by Stewart Sell (Humana Press, 2003). You might also want to look at two earlier collections published by Prometheus Books in the same series as this dealing with related scientific, moral, and religious issues. These are *Cloning: Responsible Science or Technomadness?* edited by Michael Ruse and Aryne Sheppard (Prometheus Books, 2000), and *Genetically Modified Foods: Debating Biotechnology*, edited by Michael Ruse and David Castle (Prometheus Books, 2002).

Glossary

Definitions with a "*" are found in the National Bioethics Advisory Commission report, *Ethical Issues in Human Stem Cell Research*, September 1999.

adult stem (AS) cells:* stem cells found in the adult organism (e.g., bone marrow, skin, intestine) that replenish tissues in what cells often have limited life spans. They are more differentiated than embryonic stem (ES) cells or embryonic germ (EG) cells.

a priori: a claim as in logic and pure mathematics, the truth of which is not dependent on experiment or sense experience.

amino acids: the building blocks of proteins.

antibodies: proteins that bind to foreign chemicals or microorganisms (antigens) to make them harmless; an important component of the immune system.

ART (assisted reproductive technology):* all treatments or procedures that involve handling human eggs and sperm for the purpose of helping a woman become pregnant. Types of ART include in vitro fertilization, gamete infrafallopian transfer, zygote intrafallopian transfer, embryo cryopreservation, egg or embryo donation, and surrogate birth.

artificial insemination (AI): a process by which sperm are collected from males and deposited in the females by instruments.

artificial selection: selection by humans and breeding of those individual plants or animals that possess desirable traits.

asexual reproduction: reproduction where there is no fusion of gametes and in which the genetic makeup of parents and offspring is usually identical.

bioethics: a discipline concerned with the application of ethics to biological problems, especially in the field of medicine.

bioreactors: a containment vessel for biological reactions, used in particular for fermentation processes and enzymatic reactions.

biotechnology: technological processes (agriculture, manufacturing, medicine) that involve the use of biological systems.

blastocyst:* the blastula stage in the development of the mammalian embryo (* a mammalian embryo in the state of development that follows the morula. It consists of an outer layer or trophoblast to which is attached an inner cell mass).

blastomere:* one cell of the blastula; a cell formed during early cleavage of a vertebrate embryo (* one of the cells into which the egg divides after it is fertilized; one of the cells resulting from the division of a fertilized ovum).

blastomere separation: the process in which the cells in blastomeres are made to separate; cloning through blastomere separation yields embryos with identical sets of nuclear DNA *and* mitochondrial DNA; also called embryo splitting.

blastula: usually a spherical structure produced by cleavage of a fertilized egg cell; consists of a single layer of cells surrounding a fluid-filled cavity called the blastocoel.

blood plasma: the fluid portion of blood containing proteins, salts, and other substances, but excluding the blood cells.

carcinogen: a chemical that causes cancer, generally by altering the structure of DNA.

cell cleavage: first of several cell divisions in early embryonic development that converts the zygote into a multicellular embryo.

cell culture: growth of tissue cells in artificial media.

cell cycle: the series of events in the life of a cell consisting of mitosis or nuclear division, cytokenesis or cytoplasmic division, and the interphase during which typically cells grow.

cell division: the process by which cells multiply.

cellular differentiation: see **differentiation**.

chimera:* on organism composed of two genetically distinct types of cells.

chromatin: the complex of DNA, protein, and RNA, that makes up chromosomes.

chromosome: a subcellular structure containing DNA plus the proteins that are associated with the DNA.

clone: multiple copies of identical DNA sequences that are produced when they are inserted into cloning vehicles (plasmids and other vectors) and replicated in bacteria using these cloning vehicles.

cloning:* the production of a number of genetically identical DNA molecules by inserting the chosen DNA into a phage or plasmid vector; the vector then being used to infect a suitable host within which the chosen DNA is replicated to form a number of copies, or clones (* production of a precise genetic copy of a molecule [including DNA], cell, tissue, plant, or animal).

conjoined twins: offspring that are physically attached at birth; occurs when the early embryo subdivides forming two groups of cells that do not separate completely; also referred to as Siamese twins.

consequentialist ethics: morally right actions are those that maximize the good. See **utilitarianism**.

cytoplasm: general cellular contents exclusive of the nucleus.

dedifferentiation: the opposite process of differentiation where cells or structures revert back to a less specialized stage.

deontological ethics: a moral theory according to which the rightness or necessity of an action is not exclusively determined by the value of its consequences.

developmental biology: the study of the processes by which cells and cell structures typically become more complex.

differentiated cell: a cell that has undergone the process of cellular differentiation and typically becomes specialized, e.g., a muscle cell.

differentiation:* a process changing a young, relatively unspecialized cell to a more specialized cell (* specialization of characteristics or functions of cell types).

diploid:* the condition of having a paired set of chromosomes (* cell containing two complete sets of genes derived from the father and mother, respectively; normal chromosome complement of somatic cells; in humans, forty-six chromosomes).

dizygotic twins: offspring that develop when two eggs are fertilized by a different sperm; also called fraternal twins.

DNA (deoxyribonucleic acid): a long, chainlike molecule that transmits genetic information.

***E. coli* (*Escherichia coli*):** a bacterium whose natural habitat is the gut of humans and other warm-blooded animals.

ectoderm:* the outer of the three tissue layers of the early animal embryo; typically gives rise to the skin and nervous systems; see also **mesoderm** and **endoderm** (* outer layer of cells in the embryo; origin of skin, pituitary gland, mammary glands, and all parts of the nervous system).

embryo:* a young multicellular organism that develops from the zygote and has not yet become free-living; the developing human organism until the end of the second month, after which it is referred to as a fetus (* [1] beginning of any organism in the early stages of development, [2] a state (between ovum and fetus) in the prenatal development of a mammal, [3] in humans, the state of development between the second and eighth weeks after fertilization, inclusive).

embryo splitting: see **blastomere separation.**

embryology: the scientific study of the development of organisms during the embryonic stage.

embryonic stem (ES) cell: a formative cell capable of giving rise to all other kinds of cells.

embryonic stem (ES) cells:* cells that are derived from the inner cell mass of a blastocyst embryo.

embryonic germ (EG) cells:* cells that are derived from precursors of germ cells from a fetus.

end-in-itself: a Kantian concept that something has value independent of external material ends.

endoderm:* the inner layer of the three basic tissue layers that appear during early development; becomes the digestive tract and its outgrowths—the liver, lungs, and pancreas; see also **ectoderm** and **mesoderm** (* innermost of the three primary layers of the embryo; origins of the digestive tract, liver, pancreas, and lining of the lungs).

enucleated cell or egg: a cell or egg where the nucleus has been removed.

enucleation: the process of removing the nucleus from a cell or egg.

enzyme: a protein that speeds or slows a specific chemical reaction.

epigenesis: the theory that an embryo develops from a structureless cell by the gradual addition of new parts.

epigenetic change: any change of gene activity during the development of the organism from the fertilized egg to the adult.

epithelial cells: cells that make up the tissue that covers body surfaces, lines body cavities, and forms glands.
eugenics: the claim that the way to improve humankind is through selective breeding.
ex utero:* outside of the uterus.
fertility: the state or capacity for producing offspring.
fertilization: the fusion of male and female gametes resulting in the formation of a zygote.
fetus, fetal: an unborn offspring in a late stage of development; from the third month of pregnancy to birth in humans.
fibroblasts:* connective tissue cells that secrete the intercellular material of, for example, bones and cartilage (* a cell present in connective tissue, capable of forming collagen fibers).
gamete:* a sex cell; in plants and animals, an egg or sperm (* [1] any germ cell, whether ovum or spermatozoon, [2] a mature male or female reproductive cell).
gastrula: early stage of development that follows the blastula stage.
gtastrulation:* process of transforming the blastula into the gastrula, at which point embryonic germ layers or structures begin to be laid out.
gene: the unit of heredity that consists of a stretch of DNA; contains information for the construction of other molecules, mostly proteins.
gene therapy: the use of DNA sequences either to switch off defective genes or to insert genes that specify the correct product.
genetic determinism: the claim that the traits of organisms are a direct function of their genes, not subject to environmental influences.
genetic diversity: a measure of the differences in allelic frequencies between isolated or semi-isolated populations due to various evolutionary forces such as selection and genetic drift.
genetic engineering: the manipulation of the DNA content of an organism to alter the characteristics of that organism.
genetic modification (GM): a process (especially an artificial process) that results in a change in the genetic makeup of a population.
genetic selection: artificial selection of specific genes.
genome: the genetic material of an organism.
genotype: the organism's genetic information, as distinguished from its physical appearance (phenotype).

germ cell:* cells within the body that give rise to gametes (* gametes [ova and sperm] or cells that give rise directly to gametes).

gestation: the duration of pregnancy.

G0 phase: phase in the mitotic cycle; typically a stage of arrested cellular activity.

growth phase: see **interphase**.

haploid:* the condition of having a single set of chromosomes, received from father or mother (* a cell with half the number of chromosomes as the somatic diploid cell, such as the ova or sperm. In humans, the haploid cell contains twenty-three chromosomes).

histocompatibility antigens: systems of allelic alloantigens that can stimulate an immune response leading to transplant rejection when the donor and recipient are mismatched.

hormones: "messenger" molecules that coordinate the actions of various tissues of the body; hormones produce a specific effect on the activity of target cells remote from their point of origin.

identical twins: see **monozygotic twins**.

immunology: the study of specific defense mechanisms (immune responses).

implantation: the attachment of the developing embryo to the uterine wall.

infertility: the inability of a couple to achieve conception (fertilization and establishment of pregnancy) through sexual intercourse

interphase: phase of the cell cycle during which growth occurs.

intracytoplasmic sperm injection: the process by which sperm cells are inserted directly into an egg cell.

in utero: occurring in the uterus.

in vitro:* occurring outside a living organism (literally "in glass") (* in an artificial environment, such as a test tube or culture medium).

in vivo:* occurring in a living organism (* in the natural environment, i.e., within the body).

IVF (in vitro fertilization):* the process during which an egg is removed from the ovary and fertilized with sperm in laboratory glassware and then reimplanted in the female (* a process by which a woman's eggs are extracted and fertilized in the laboratory and transferred after they reach the embryonic stage into the woman's uterus through the cervix. Roughly 70 percent of assisted reproduction attempts involve IVF using fresh embryos developed from a woman's own eggs).

Kantian ethics: based on the categorical imperative, namely that we should act only according to those maxims that can be consistently willed as a universal law.

karyotype:* chromosome characteristics of an individual cell or cell line, usually presented as a systematic array of metaphase chromosomes from a photograph of a single cell nucleus arranged in pairs in descending order of size.

keratinocyte: a specialized epidermal cell that synthesizes keratin, the substance that makes up nails and feathers.

lactation: the production or release of milk from the breast.

mammary gland: the organs that secrete milk to feed offspring.

meiosis: process in which a diploid cell undergoes two successive nuclear divisions to produce four haploid cells; the process that produces gametes in animals.

mesoderm:* the middle layer of the three basic tissue layers that develop in the early embryo; gives rise to connective tissue, muscle, bone, blood vessels, and many other structures; see also **ectoderm** and **endoderm** (* middle of the three primary germ layers of the embryo; origin of all connective tissues, all body musculature, blood, cardiovascular and lymphatic systems, most of the urogenital system, and lining of the pericardial, pleural, and peritoneal cavities).

metabolism: a collective term for the chemical processes involved in the maintenance of life and by which energy is made available.

metaphase: the stage of mitosis and meiosis during which the chromosomes line up on the equatorial plane of the cell.

metaphysics: the branch of philosophy which deals with the ultimate nature and origin of things.

microinjection: injection of cells with solutions by using a micropipette.

miscarriage: the expulsion of the embryo or fetus from the uterus before it is viable.

mitochondria (sing. **mitochondrion):** a specialized subcellular structure that converts chemical energy from one form to another.

mitochondrial disease: diseases that are due to mutations or malfunctions in mitochondria.

mitochondrial DNA (mtDNA): the DNA contained in the mitochondria

mitosis: division of the cell nucleus, resulting in two daughter nuclei with the same number of chromosomes as the parent nucleus; consists of four phases: prophase, metaphase, anaphase, and telophase.

molecular biology: the chemistry and physics of the molecules that constitute living things.

monoclonal antibody: an antibody produced by a clone of antibody-producing cells that recognizes only one feature.

monovular twins: see **monozygotic twins.**

monozygotic twins: offspring that possess identical sets of genetic material; result from the division of the early embryo; also referred to as identical or monovular twins.

moral agency: the capability of having rights as an individual.

morphology: the scientific study of organic form.

morula:* (1) mass of blastomeres resulting form early cleavage divisions of the zygote, (2) solid mass of cells resembling a mulberry, resulting form cleavage of an ovum.

multipotent: see **totipotent**.

mutagen: any agent that is capable of producing changes in the DNA.

mutations: any change in the DNA , including a change in the sequence of nucleotide pairs that constitute DNA or a rearrangement of genes within chromosomes.

mutatis mutandis: after the necessary or appropriate changes have been made.

myoepithelial cells: epithelial cells of coelenterata that are provided with basal contractile outgrowths.

natural selection: the key Darwinian mechanism for evolutionary change, claiming that only a percentage of organisms in each generation survive and reproduce, and that these do so because they have characteristics which other members of the population do not possess; these "adaptive" characteristics are passed along to offspring.

nuclear: pertaining to the nucleus.

nuclear DNA: the DNA contained within the nucleus of the cell (as opposed to the DNA contained in the mitochondria).

nuclear substitution: see **nuclear transfer**.

nuclear transfer: process where the nucleus of a host cell is removed and replaced with the nucleus of a donor cell that is to be cloned

nuclear transplantation: see **nuclear transfer**.

nucleolus (pl. **nucleoli**)**:** small nuclear body made up of protein and ribosomal RNA.

nucleus (pl. **nuclei**)**:** a double membrane-bounded cellular structure containing DNA.

neurula: the stage of development of the vertebrate embryo that is characterized by the development of the neural tube from the neural plate.

nullipotent: unable to direct the development of cells unlike themselves.

oestrus: the recurrent period of heat or sexual receptivity occurring around ovulation in female mammals.

oncogenesis: the processes of tumor formation.

oocytes:* cells that give rise to egg cells (ova) by meiosis (* [1] diploid cell that will undergo meiosis (type of cell division of germ cells) to form an egg, [2] immature ovum).

ovulation: the release of a mature egg from the ovary.

ovum:* female reproductive or germ cell.

parthenogenesis: a form of asexual reproduction in which an unfertilized egg develops into an adult organism.

parturition: the process of birth.

pathogen: a disease-causing agent.

perinatal: surrounding the time of birth.

pharmaceuticals: chemicals used in medicine.

phenomenology: the philosophy in which supposedly we come to know mind as it is in itself through the study of the way in which it appears to us.

phenotype: the physical and behavioural characteristics of an organism.

physiology: the scientific study of the workings of living bodies and their parts.

ploidy: relating to the number of sets of chromosomes in a cell.

pluripotent:* capable of giving rise to various kinds of embryonic tissue (* cells, present in the early stages of embryo development, that can generate all cell types in a fetus and in the adult and that are capable of self-renewal. Pluripotent cells are not capable of developing into an entire organism).

polyclonal antibody: an antibody produced by many clones of cells that recognize several molecular features of an antigen.

polymorphic loci: genetic loci, in a population, which have several different alleles.

post mortem: the period after death; also, the examination of an organism after death.

pre-embryos: the preimplantation stages of the development of the zygote, occurring during the first two weeks after fertilization.

preformation theory: the theory of early developmental biologists that the fully formed animal or plant exists in a minute form in the germ cell; see **epigenesis.**

preimplantation diagnosis: the determination of the genotype of an in vitro–fertilized human embryo (for genetic diseases) prior to its implantation.

preimplantation embryo:* (1) embryo before it has implanted in the uterus, (2) commonly used to refer to in vitro–fertilized embryos before they are transferred to a woman's uterus.

prenatal diagnosis (or screening): fetal cells are sampled and analyzed for chromosomal and biochemical disorders.

prima facie: clear and apparent at first glance, presumed true unless disproven by subsequently acquired evidence.

progenitors: the ancestors of an organism.

progeny: offspring.

promoter region: a site on the DNA that facilitates the initiation of transcription.

pronucleus (pl. pronuclei): one of the two nuclear bodies of a newly fertilized egg cell; the male and the female pronuclei fuse to form the nucleus of the new cell.

quiescence: the state of dormancy or inactivity.

recessive gene: any gene whose expression is masked by another (dominant) gene.

recombinant DNA: hybrid DNA produced by joining pieces of DNA from different sources.

reductio ad absurdum: refutation of an assumption by deriving a contradiction (or otherwise clearly false conclusion) from the original assumption.

reproductive cloning: the use of cloning technology to produce new individuals (in contrast to therapeutic cloning).

senescence: the aging process.

sexual reproduction: form of reproduction in which two gametes fuse to form a zygote.

Siamese twins: see **conjoined twins.**

slippery slope: an argument that an action apparently unobjectionable in itself would set in motion a train of events leading ultimately to an undesirable outcome.

somatic cell:* body cell, as distinguished from a germ cell (* [from *soma,* the body] (1) cells of the body which in mammals and flowering plants normally are made up of two sets of chromosomes, one derived from each parent, (2) all cells of an organism with the exception of germ cells).

somatic cell nuclear transfer (SCNT): the process by which a nucleus from a (somatic) body cell is removed and inserted into an egg cell.

spermatogonium: germ cell produced by a male.

stem cell:* see **embryonic stem (ES) cell** (* cells that have the ability to divide indefinitely and to give rise to specialized cells as well as to new stem cells with identical potential).

surrogate: a female whose uterus is used to gestate an embryo that is not her own.

teleology: a theory that describes or explains in terms of purpose.

telomerase: the enzyme that elongates telomeres.

telomere: the end of a chromosome in which the DNA is looped back.

telos: endpoint, aim, goal, or purpose of an activity, or a final state of affairs.

therapeutic cloning: the use of cloning technology for medical purposes, for example, the production of organs for transplantation (in contrast to reproductive cloning).

tissue culturing: a method of propagating healthy cells outside the organism.

totipotency (adj. **totipotent):*** ability of a cell (or nucleus) to develop into any other type of cell (or nucleus) (* having unlimited capacity. Totipotent cells have the capacity to differentiate into the embryo and into extraembryonic membranes and tissues. Totipotent cells contribute to every cell type of the adult organism).

transcription: the process of converting information in DNA into information in RNA (ribonucleic acid); RNA functions mainly in protein synthesis.

transgene: a gene that is taken from one organism and inserted into the genome of another.

transgenesis: genetic modification of animals using transgenes.

transgenic animal: an animal that has incorporated foreign DNA into its genome.

translation: the process of converting the information in RNA into protein; also called protein synthesis.

trimester: one of the three three-month periods of human pregnancy

trophoblast:* outermost layer of the developing blastocyst of a mammal. It differentiates into two layers, the cytotrophoblast and syntrophoblast, the latter coming into intimate relationship with uterine endometrium with which it establishes a nutrient relationship.

undifferentiated cell: an unspecialized cell that has not undergone differentiation.

unfertilized egg: an egg cell that has not undergone fertilization.

utilitarianism: the moral theory that an action is morally right if and only if it produces at least as much good (utility) for all people affected by the action as any alternative action.

vector: an agent, such as a bacterial plasmid or virus, that transfers genetic information.

xenotransplantation: transplantation of tissue or organs between species.

zygote:* one-celled embryo formed by the fusion of two germ cells (* [1] cell resulting from fusion of two gametes in sexual reproduction, [2] fertilized egg (ovum), [3] diploid cell resulting from union of sperm and ovum, [4] developing organism during the first week after fertilization).

Contributors

Marcia Barinaga is a contributing correspondent (Berkeley Bureau) for *Science*. She writes about neuroscience and other areas of basic biology; also writes policy and other stories with a West Coast angle.

George W. Bush is the forty-third president of the United States. Formerly the forty-sixth governor of the state of Texas. He is the son of the forty-first president of the United States, George Herbert Walker Bush; and he is the second son of a former president to become president after John Quincy Adams, son of John Adams, was elected in 1824. Bush was reelected in November 2004.

Arthur Caplan has been the director of the Center for Bioethics and trustee professor of Bioethics at the University of Pennsylvania since 1994. He is currently chairman of the advisory committee to the Department of Health and Human Services, Centers for Disease Control, and Food and Drug Administration on Blood Safety and Availability.

Simon Clarke received his D. Phil. from the University of Oxford in 2001 and before that did his BA and MA at the University of Auckland. He also spent one semester at Columbia University in New York City as a visiting scholar. His areas of interest are political philosophy, the history of political thought, ethics, and legal philosophy. As well as tutoring at Oxford and Auckland universities, he lectured in the Political Science Department at Canterbury before joining the Philosophy Department.

Maureen L. Condic is an assistant professor of neurobiology and anatomy at the University of Utah School of Medicine. Recent work from her laboratory has demonstrated that the regeneration of adult neurons in culture can be greatly improved by transgenic integrin expression. She is currently expanding these findings to more sophisticated in vitro and whole-animal models of adult regeneration.

Ruth Faden and the Rest is the Philip Franklin Wagley Professor of Biomedical Ethics and director of the Bioethics Institute, Johns Hopkins University. She is also a senior research scholar at the Kennedy Institute of Ethics, Georgetown University. Dr. Faden is the author and editor of numerous books and articles on biomedical ethics and health policy including *A History and Theory of Informed Consent* (with Tom L. Beauchamp), *AIDS, Women and the Next Generation* (Ruth Faden, Gail Geller, and Madison Powers, eds.) and *HIV, AIDS and Childbearing: Public Policy, Private Lives* (Ruth Faden and Nancy Kass, eds.). Dr. Faden is a member of the Institute of Medicine and a fellow of the Hastings Center and the American Psychological Association. She has served on several national advisory committees and commissions. Most recently, she was the chair of the President's Advisory Committee on Human Radiation Experiments.

Geron Ethics Board, an ethics advisory board, whose members represent a variety of philosophical and theological traditions with a breadth of experience in health care ethics, was created by Geron Corporation in 1998. The board functions as an independent entity, consulting and giving advice to the corporation on ethical aspects of the work Geron sponsors. Members of the board have no financial interest in Geron Corporation.

Sidney Houff is professor and chairman of the Department of Neurology at the Loyola University Health Systems at Loyola University, Chicago. He is most recognized as a researcher who has studied AIDS-related neurological diseases and was the first scientist to report human-to-human transmission of rabies via corneal transplants. Dr. Houff plans on continuing research in how viruses alter areas of the brain responsible for behavior and cognitive function and as chairman will expand the comprehensive epilepsy program, multidisciplinary stroke center, and broaden neurology services to pediatrics. Long-term goals include further integration of clinical and basic sciences in the study of a number of neurological diseases.

Aline H. Kalbian is an assistant professor of religion at Florida State University. She teaches in the area of religious ethics and gender and ethics. Her areas of interest include contemporary Catholic moral theology, gender and ethics, medical ethics, and sexual ethics. She is presently working on the book *Controlling Reproduction/Reproducing Gender: The Catholic Church and Human Sexuality*. Her other research interests include ethics and reproductive technology and gender metaphors in recent Catholic magisterial writings.

Katty Kay and Mark Henderson are reporters for the *Times* in London.

Søren Holm, CCELS's director, joined the Cardiff Law School in 2004 to take up the chair in bioethics at the Cardiff Institute for Society, Health, and Ethics. Previously professor of bioethics at the Law School in Manchester, he is also a former member of the Danish Council of Ethics. He has published widely on issues in both philosophy of medicine and health care ethics, and has research collaborations with individuals and centers in many European countries. Since May 2004 he is joint editor in chief of the *Journal of Medical Ethics*.

Jane Maienschein is a professor of philosophy and biology at Arizona State University in Tempe, Arizona. Dr. Maienschein specializes in the history and philosophy of biology. Looking at research in embryology, genetics, and cytology in the first half of this century, she asks what factors have brought changes in biological theory and practice. She has concluded that changing epistemological convictions about what counted as good science caused researchers to address increasingly narrower and more specialized questions and to look for concrete empirical results. Biologists began to accept that they were generating temporary working hypotheses rather than always finding the "truth," but they regarded their "facts" as definitive. As the science became more technically complex, a concept of "cutting-edge" science emerged and began to define success in biology. Dr. Maienschein combines detailed analysis of the epistemological standards, theories, laboratory practices, organisms selected, techniques, and equipment with study of the people, institutions, and their historical settings to explore the changing nature of the biological sciences and the social, political, and legal context in which science thrives.

Carol Marin was named a correspondent for CBS News in October 2000. In this role, Marin primarily serves as a contributor to *60 Minutes* and *60 Minutes II*.

Don Marquis is professor of philosophy at the University of Kansas, Lawrence. He is the author of the often reprinted "Why Abortion Is Immoral," which presents a secular argument that abortion is immoral. That paper was originally published in the *Journal of Philosophy* in 1989.

Glenn McGee is an associate director at the Center for Bioethics of the University of Pennsylvania, where he holds appointments in philosophy, history, and sociology of science, and cellular engineering. He is editor in chief of the *American Journal of Bioethics* and author of a number of articles about ethical issues in genetics, stem cell research, cloning, and reproduction.

National Bioethics Advisory Commission (NBAC) was created by President William Jefferson Clinton with Executive Order 12975 on October 3, 1995. The purpose of the NBAC was to provide a review of policies and procedures in the area of bioethics. The commission has produced reports on the subjects of cloning and stem cell research. Its was directed by Harold T. Shapiro, and the NBAC's charter expired on October 3, 2001.

National Institutes of Health (NIH) is one of eight health agencies of the Public Health Services which, in turn, is part of the U.S. Department of Health and Human Services. Comprised of twenty-seven separate components, mainly institutes and centers, NIH has seventy-five buildings on more than three hundred acres in Bethesda, Maryland. From a total of about $300 in 1887, the NIH budget has grown to more than $20.3 billion in 2001.

Michael Novak is the George Frederick Jewett Scholar in Religion, Philosophy, and Public Policy Director of Social and Political Studies at the American Enterprise Institute in Washington, D.C. He received the Templeton Prize for Progress in Religion (a million-dollar purse awarded at Buckingham Palace) in 1994, and delivered the Templeton address in Westminster Abbey. He is a frequent contributor to *National Review*, *First Things*, and *New Republic*.

Daniel Perry is executive director of the Alliance for Aging Research (www.agingresearch.org), a Washington, D.C.–based nonprofit organization dedicated to improving the health and independence of aging Americans through public and private funding of medical research and geriatric education.

Ted Peters is professor of systematic theology at Pacific Lutheran Theological Seminary and the Graduate Theological Union, Berkeley, California, and a research associate at the Center for Theology and the Natural Sciences. He also serves on the GERON ethics board.

Christopher A. Pynes is assistant professor of philosophy at Western Illinois University in the Department of Philosophy and Religious Studies. He has taught at the University of Tennessee, Knoxville, and at Florida State where he received his Ph.D. His philosophical interests are in ethics, logic, philosophy of science, and analytic jurisprudence. He was also coeditor of the first edition of *The Stem Cell Controversy* with Michael Ruse.

John A. Robertson holds the Vinson and Elkins Chair at the University of Texas School of Law at Austin and currently serves as cochair of the ethics committee of the American Society for Reproductive Medicine.

Michael Ruse is Lucyle T. Werkmeister Professor of Philosophy at Florida State University. He began his teaching career at the University of Guelph, where he taught for thirty years. He is a fellow of the Royal Society of Canada. He writes on the nature of science, in particular evolutionary biology, and the nature of value. A list of his most recent books would include the following: *Can a Darwinian Be a Christian?*; *The Relation between Science and Religion*; *Mystery of Mysteries: Is Evolution a Social Construction?*; *Monad to Man: The Concept of Progress in Evolutionary Biology*; and *Darwin and Design: Does Evolution Have a Purpose?*

Kenneth J. Ryan is the Kate Macy Ladd Distinguished Professor of Obstetrics, Gynecology, and Reproductive Biology, Emeritus at Harvard University. From 1973 to 1993, Dr. Ryan served as chairman of the Department of Obstetrics, Gynecology, and Reproductive Biology at Harvard and the Brigham & Women's Hospital. From 1974 to 1978, he served as chairman of the

National Commission for the Protection of Human Subjects of Biomedical and Behavioral Research. Dr. Ryan is former chairman of the ethics committee of the Brigham & Women's Hospital and serves on many professional and governmental committees and panels related to ethics and public policy in medicine and human subjects research. He has received numerous awards and fellowships and published in the areas of endocrinology, human reproduction, and medical ethics.

Abdulaziz Sachedina was born in Tanzania, and is professor of religious studies at the University of Virginia. His Ph.D. is from the University of Toronto, and he has a BA from Aligarh Muslim University in India and Ferdowsi University in Iran. He has been visiting professor at Wilfrid Laurier, Waterloo, and McGill Universities in Canada, Haverford College and the University of Jordan, Amman. He has lectured widely in East Africa, India, Pakistan, Europe, and the Middle East. Professor Sachedina is a core member of the "Islamic Roots of Democratic Pluralism" project in the CSIS Preventive Diplomacy program and a key contributor to the program's efforts to link religion to universal human needs and values in the service of peace-building.

Leigh Shoemaker is a Ph.D. student in philosophy at the University of Tennessee, Knoxville. She holds masters' degrees in both philosophy and information science. Her primary areas of interest within philosophy are environmental ethics, philosophy of science, autonomy, and epistemology.

Andrew W. Siegel is associate director of academic programs at the Bioethics Institute. He received his J.D. and Ph.D. in philosophy from the University of Wisconsin-Madison. He was a Greenwall Fellow at Johns Hopkins and Georgetown, 1996–1998. Dr. Siegel has served as staff philosopher for the National Bioethics Advisory Commission, legislative fellow for Senator Edward M. Kennedy and the Labor and Human Resources Committee, and staff attorney for the Task Force on Genetic Testing of the Working Group on the Ethical, Legal, and Social Issues of the Human Genome Project. He has taught philosophy at Georgetown and Johns Hopkins, and law at Georgetown and the University of Wisconsin. His current research focuses on ethical and legal issues in human stem cell research and end-of-life decision making, as well as on more theoretical problems in political philosophy and philosophy of mind.

Cindy R. Towns and D. Gareth Jones work at the University of Otago in New Zealand.

Gretchen Vogel is the Berlin correspondent for *Science*. Her areas of interest are science policy in Germany and Central Europe, developmental biology, and evolution.

LeRoy Walters attended a small Pennsylvania liberal-arts college, Messiah College, receiving his B.A. in 1962. In 1967 he began a Ph.D. program in the Department of Religious Studies at Yale University where he eventually earned the degree. During the summer of 1996, Professor Walters accepted a three-year term as director of the Kennedy Institute. He has also served for three terms on the Recombinant DNA Advisory Committee of the National Institutes of Health. From 1993 through 1996 he served as chair of the committee, which reviews human-gene-therapy protocols. A continuing interest in Professor Walters's work has been the development of a bioethics library. The Kennedy Institute library, now called the National Reference Center for Bioethics Literature, is the largest collection of materials on biomedical ethics under one roof in the world.

Rick Weiss is a science and medical reporter for the *Washington Post*. He came to the *Post*'s health section in 1993 and moved to the national desk in January 1996, where he covers genetics, molecular biology, and other topics in the life sciences.

Frank E. Young has served in a variety of governmental positions, including commissioner of the Food and Drug Administration (1984–1989), commissioner on the World Health Organization's Commission on Health and Environment (1990), and assistant surgeon general (rear admiral, upper half) of the Department of Health and Human Services (1984). He is the former dean of the School of Medicine and Dentistry at the University of Rochester and is an ordained minister in the Evangelical Church Alliance.